Digital Photogrammetry

Photogrammetry, the use of photography for surveying, primarily facilitates the production of maps and geographic databases from aerial photographs. Along with remote sensing, it represents the principal means of generating data for geographic information systems.

Photogrammetry has undergone a remarkable evolution in recent years with its transformation into 'digital photogrammetry'. First, the distinctions between photogrammetry, remote sensing, geodesy and GIS are fast disappearing, as data can now be carried digitally from the plane to the GIS end-user. And second, the benefits of digital photogrammetric workstations have increased dramatically. The comprehensive use of digital tools, and the automation of the processes, have significantly cut costs and reduced processing time. The first digital aerial cameras have become available, and the introduction of more and more new digital tools allows the work of operators to be simplified, without the same need for stereoscopic skills. Engineers and technicians in other fields are now able to carry out photogrammetric work without reliance on specialist photogrammetrists.

This book shows non-experts what digital photogrammetry is and what the software does, and provides them with sufficient expertise to use it. It also gives specialists an overview of these totally digital processes from A to Z. It serves as a textbook for graduate students, young engineers and university lecturers to complement a modern lecture course on photogrammetry.

Michel Kasser and Yves Egels have synthesized the contributions of 21 top-ranking specialists in digital photogrammetry, lecturers, researchers, and production engineers, and produced a very up-to-date text and guide.

Michel Kasser, who graduated from the École Polytechnique de Paris and from the École Nationale des Sciences Géographiques (ENSG), was formerly Head of ESGT, the main technical university for surveyors in France. A specialist in instrumentation and space imagery, he is now University Professor and Head of the Geodetic Department at IGN-France.

Yves Egels, who graduated from the École Polytechnique de Paris and from ENSG, is senior photogrammetrist at IGN-France, where he pioneered analytical plotter software in the 1980s and digital workstations on PCs in the 1990s. He is now Head of the Photogrammetric Department at ENSG and lectures in photogrammetry at various French technical universities.

Digital Photogrammetry

Michel Kasser and Yves Egels

London and New York

First published 2002
by Taylor & Francis
11 New Fetter Lane, London EC4P 4EE

Simultaneously published in the USA and Canada
by Taylor & Francis Inc,
29 West 35th Street, New York, NY 10001

Taylor & Francis is an imprint of the Taylor & Francis Group

© 2002 Michel Kasser and Yves Egels

Typeset in Sabon by
Florence Production Ltd, Stoodleigh, Devon
Printed and bound in Great Britain by
Biddles Ltd, Guildford and King's Lynn

All rights reserved. No part of this book may be reprinted or reproduced or utilised in any form or by any electronic, mechanical, or other means, now known or hereafter invented, including photocopying and recording, or in any information storage or retrieval system, without permission in writing from the publishers.

Every effort has been made to ensure that the advice and information in this book is true and accurate at the time of going to press. However, neither the publisher nor the authors can accept any legal responsibility or liability for any errors or omissions that may be made. In the case of drug administration, any medical procedure or the use of technical equipment mentioned within this book, you are strongly advised to consult the manufacturer's guidelines.

British Library Cataloguing in Publication Data
A catalogue record for this book is available from the
British Library

Library of Congress Cataloging in Publication Data
Kasser, Michel
  Digital photogrammetry/Michel Kasser and Yves Egels.
    p. cm
  Includes bibliographical references and index.
  I. Aerial photogrammetry.   2. Image processing – Digital techniques.
  I. Kasser, Michel.   II. Title.
  TA 593.E34 2001
  526.9′82 – dc21   2001027205

ISBN 0–748–40945–9 (pbk)
ISBN 0–748–40944–0 (hbk)

Contents

List of colour plates	viii
List of contributors	ix
Introduction	xiii

1 Image acquisition. Physical aspects. Instruments   1

Introduction   1

  1.1 Mathematical model of the geometry of the aerial image   2
      YVES EGELS

  1.2 Radiometric effects of the atmosphere and the optics   16
      MICHEL KASSER

  1.3 Colorimetry   25
      YANNICK BERGIA, MICHEL KASSER

  1.4 Geometry of aerial and spatial pictures   34
      MICHEL KASSER

  1.5 Digital image acquisition with airborne CCD cameras   39
      MICHEL KASSER

  1.6 Radar images in photogrammetry   47
      LAURENT POLIDORI

  1.7 Use of airborne laser ranging systems for the determination of DSM   53
      MICHEL KASSER

  1.8 Use of scanners for the digitization of aerial pictures   58
      MICHEL KASSER

  1.9 Relations between radiometric and geometric precision in digital imagery   63
      CHRISTIAN THOM

2 Techniques for plotting digital images — 78

Introduction — 78

- 2.1 Image improvements — 79
  ALAIN DUPÉRET
- 2.2 Compression of digital images — 100
  GILLES MOURY
- 2.3 Use of GPS in photogrammetry — 115
  THIERRY DUQUESNOY, YVES EGELS, MICHEL KASSER
- 2.4 Automatization of aerotriangulation — 124
  FRANCK JUNG, FRANK FUCHS, DIDIER BOLDO
- 2.5 Digital photogrammetric workstations — 145
  RAPHAËLE HENO, YVES EGELS

3 Generation of digital terrain and surface models — 159

Introduction — 159

- 3.1 Overview of digital surface models — 159
  NICOLAS PAPARODITIS, LAURENT POLIDORI
- 3.2 DSM quality: internal and external validation — 164
  LAURENT POLIDORI
- 3.3 3D data acquisition from visible images — 168
  NICOLAS PAPARODITIS, OLIVIER DISSARD
- 3.4 From the digital surface model (DSM) to the digital terrain model (DTM) — 221
  OLIVIER JAMET
- 3.5 DSM reconstruction — 225
  GRÉGOIRE MAILLET, PATRICK JULIEN, NICOLAS PAPARODITIS
- 3.6 Extraction of characteristic lines of the relief — 253
  ALAIN DUPÉRET, OLIVIER JAMET
- 3.7 From the aerial image to orthophotography: different levels of rectification — 282
  MICHEL KASSER, LAURENT POLIDORI
- 3.8 Production of digital orthophotographies — 288
  DIDIER BOLDO
- 3.9 Problems relating to orthophotography production — 292
  DIDIER MOISSET

4 Metrologic applications of digital photogrammetry   300
Introduction   300
4.1 Architectural photogrammetry   300
PIERRE GRUSSENMEYER, KLAUS HANKE, ANDRÉ STREILEIN

4.2 Photogrammetric metrology   340
MICHEL KASSER

Index   349

Plates

There is a colour plate section between pages 208 and 209

Figure 3.3.20 Left and right images of a 1m ground pixel size satellite across track stereopair; DSM result with a classical cross-correlation technique; DSM using adaptive shape windows

Figure 3.3.24 An example of the management of hidden parts

Figure 3.3.27 Correlation matrix corresponding to the matching of the two epipolar lines appearing in yellow in the image extracts

Figure 3.3.28 Results of the global matching strategy on a complete stereopair overlap in a dense urban area

Figure 3.4.1 Images from the digital camera of IGN-F (1997) on Le Mans (France)

Figure 3.8.3 Scanned images before balancing

Figure 3.8.4 Scanned images after balancing

Figure 3.9.2 Example from a real production problem: the radiometric balancing on digitized pictures

Figure 3.9.4 Examples from the Ariège, France: problems posed by cliffs

Figure 3.9.5 Generation of stretched pixels

Figure 3.9.6 Example of radiometric difficulties (Ariège, France)

Figure 3.9.7 Example of typical problems on water surfaces

Figure 3.9.8 Two examples from the Isère department, France

Figure 3.9.9 Two examples, using the digital aerial camera of IGN-F

Contributors

Yannick Bergia, born in 1977, graduated as an informatics engineer from ISIMA (Institut Supérieur d'Informatique, de Modélisation et de leurs Applications, Clermont-Ferrand, France) in 1999. He performed his diploma work in MATIS in 1999, and is a specialist in image processing.
y.bergia@infonie.fr

Didier Boldo, born in 1973, graduated from the École Polytechnique de Paris and from ENSG. He is now pursuing a Ph.D. in pattern recognition and image analysis at MATIS. His main research interests are photogrammetry, 3D reconstruction and radiometric modelling and calibration.
didier.boldo@ign.fr

Olivier Dissard, born in 1967, graduated from the École Polytechnique de Paris and from ENSG. Previously in charge of research concerning 3D urban topics at MATIS laboratory, he is now in charge of digital orthophotography at IGN-F. His studies at MATIS have concerned urban DEM and DSM, focusing on raised structures and classification in buildings and vegetation, and true orthophotographies.
olivier.dissard@ign.fr

Alain Dupéret graduated as ingénieur géographe from ENSG. He started working as a surveyor in IGN-F in 1979 and undertook topographic missions in France and Africa before specialising in image processing and DTM production. After managing technical projects and giving lectures in photogrammetry and image processing in ENSG, he became Head of Studies Management at ENSG.
duperet@ensg.ign.fr

Thierry Duquesnoy (born in 1963) graduated from ENSG in 1989. He gained his Ph.D. in earth sciences in 1997 at the University of Paris. He has worked since 1989 with the LOEMI, where he is a specialist in GPS trajectography for photogrammetry.
thierry.duquesnoy@ign.fr

Contributors

Yves Egels, born in 1947, graduated from the École Polytechnique de Paris and from ENSG, and is senior photogrammetrist at IGN-F, where he pioneered analytical plotter software in the 1980s and digital workstations on PCs in the 1990s. He is now Head of the Photogrammetric Department at ENSG and gives lectures in photogrammetry at various French technical universities.
egels@ensg.ign.fr

Frank Fuchs, born in 1971, graduated from the École Polytechnique de Paris and from ENSG. Since 1996 he has been working at MATIS on his Ph.D. concerning the automatic reconstruction of buildings in aerial imagery through a structural approach.
frank.fuchs@ign.fr

Pierre Grussenmeyer, born in 1961, graduated from ENSAIS, and gained a Ph.D. in photogrammetry in 1994 at the University of Strasbourg in collaboration with IGN-F. He has been on the academic staff of the Department of Surveying at ENSAIS, where he teaches photogrammetry, since 1989. Since 1996 he has been Professor and the Head of the Photogrammetry and Geomatics Group at ENSAIS-LERGEC.
pierre.grussenmeyer@ensais.u-strasbg.fr

Klaus Hanke, born 1954, studied geodesy and photogrammetry at the University of Innsbruck and the Technical University of Graz. In 1984 he became Dr. techn., and since 1994 he has been a Professor teaching Photogrammetry and Architectural Photogrammetry at the University of Innsbruck.

Raphaële Heno, born 1970, graduated in 1993 from ENSG and from ITC (Enschede – The Netherlands). She spent four years developing digital photogrammetric tools to improve IGN-France topographic database plotting. Since 1998 she has been senior consultant for the IGN-France advisory department.
raphaele_heno@hotmail.com

Olivier Jamet, born in 1963, graduated from the École Polytechnique de Paris and from ENSG. He gained a Ph.D. in signal and image processing at the École Nationale Supérieure des Télécommunications de Paris (1998). Formerly Head of the MATIS laboratory, he is currently in charge of research in physical geodesy at LAREG, ENSG.
olivier.jamet@ign.fr

Patrick Julien graduated from the University of Paris (in mathematics) and from ENSG, and joined IGN-F in 1975. He has developed various softwares for orthophotography, digital terrain models and digital image matching. He is presently a researcher at MATIS and gives lectures at ENSG in geographic information sciences.
patrick.julien@ign.fr

Contributors xi

Franck Jung, born in 1970, graduated from the École Polytechnique de Paris and from ENSG, and is currently a researcher at the MATIS laboratory. He is preparing a Ph.D. dissertation on automatic change detection using aerial stereo pairs at two different dates. His main research interests are automatic digital photogrammetry, change detection and machine learning.
franck.jung@ign.fr

Michel Kasser, born in 1953, graduated from the École Polytechnique de Paris and from ENSG. He created the LOEMI in 1984 and was from 1991 to 1999 head of the ESGT. As a specialist in instrumentation, in geodesy and in space imagery, he is University Professor and Head of the Geodetic Department and of LAREG at IGN-F.
michel.kasser@ign.fr

Grégoire Maillet, born in 1973, graduated from ENSG in 1997. He is now a research engineer at MATIS, where he works on dynamic programming and automatic correlation for image matching.
gregoire.maillet@ign.fr

Didier Moisset, born in 1958, obtained the ENSG engineering degree in 1983. He was in charge of the aerotriangulation team at IGN in 1989, and was Professor in photogrammetry and Head of the Photogrammetric Department at the ENSG during 1993–8. He was a project manager in 1998 and developed new automatic orthophoto software for IGN. He was recently appointed as a consultant in photogrammetry at the MATIS laboratory.
didier.moisset@ign.fr

Gilles Moury, born in 1960, graduated from the École Polytechnique de Paris and from ENSAE (École Nationale Supérieure de l'Aéronautique et de l'Espace). He has been working in the field of satellite on-board data processing since 1985, within CNES. In particular, he was responsible for the development of image compression algorithms for various space missions (Spot 5, Clementine, etc.). He is now Head of the on-board data processing section and lectures in data compression at various French technical universities.
gilles.moury@cnes.fr

Nicolas Paparoditis, born in 1969, obtained a Ph.D. in computer vision from the University of Nice-Sophia Antipolis. He worked for five years for the Aérospatiale Company on the specification of HRV digital mapping satellite instruments (SPOT satellites). He is a specialist in image processing, in computer vision, and in digital aerial and satellite imagery, and is currently a Professor at ESGT where he leads the research activities in digital photogrammetry. He is also a research member of MATIS at IGN-F.
nicolas.paparoditis@ign.fr

Contributors

Laurent Polidori, born in 1965, graduated from ENSG in 1987. He gained a Ph.D. in 1991 on 'DTM quality assessment for earth sciences' at Paris University. A researcher at the Aérospatiale Company (Cannes, France) on satellite imagery, he is the author of *Cartographie Radar* (Gordon and Breach, 1997). He is Head of the Remote Sensing Laboratory of IRD (Cayenne, Guyane) and Associate Professor at ESGT.
polidori@caiena.cayenne.ird.fr

André Streilein in 1990 obtained a Dip.Ing. at the Rheinische Friedrich-Wilhelms-Universität, Bonn. In 1998 he obtained a Dr. sc. tech. at the Swiss Federal Institute of Technology, Zurich. His dissertation was on 'Digitale Photogrammetrie und CAAD'. Since September 2000 he has been at the Swiss Federal Office of Topography (Bern), Section Photogrammetry and Remote Sensing. (Postal address: Bundesamt für Landestopographie, Seftigenstrasse 2, 3084 Wabern, Bern, Switzerland.)

Christian Thom, born in 1959, graduated from the École Polytechnique de Paris and from the ENSG. With a Ph.D. (1986) in robotics and signal processing, he has specialized in instrumentation. He is Head of LOEMI where he pioneered and developed the IGN program of digital aerial cameras from 1989.
christian.thom@ign.fr

Institutions

IGN-F is the Institut Géographique National, France, administration in charge of national geodesy, cartography and geographic databases. Its main installation is in Saint-Mandé, close to Paris. The MATIS is the laboratory for research in photogrammetry of IGN-F; the LOEMI is its laboratory for new instrument development; and the LAREG its laboratory for geodesy. The ENSG (École Nationale des Sciences Géographiques) is the technical university owned by IGN-F.
(Postal address: IGN, 2 Avenue Pasteur, F-94 165 Saint-Mandé Cedex, France.)

The CNES is the French Space Agency.
(Postal address: CNES, 2 Place Maurice Quentin, F-75 039 Paris Cedex 01, France.)

The ENSAIS is a technical university in Strasbourg with a section of engineers in surveying.
(Postal address: ENSAIS, 24 Boulevard de la Victoire, F-67 000, Strasbourg, France.)

The ESGT is the École Supérieure des Géomètres et Topographes (Le Mans), the technical university with the most important section of engineers in surveying in France.
(Postal address: ESGT, 1 Boulevard Pythagore, F-72 000, Le Mans, France.)

Introduction

In 1998, when Taylor & Francis pushed me into the adventure of writing this book with my colleague Yves Egels, most specialists of photogrammetry around me were aware of the lack of communication between their technical community and other specialists. Indeed, this situation was not new, as 150 years after its invention by Laussédat photogrammetry was really reserved for photogrammetrists (considering for example the extremely high costs of photogrammetric plotters, but also the necessary visual ability of the required technicians, a skill that required a long and costly training, etc.). Photogrammetry was thus the speciality of a very small club, mostly European, and the related industrial activity was heavily based on the national cartographic institutions.

As we saw it in the early 1990s, the situation has changed very rapidly with the improvement of the computational power available on cheap machines. First, one could see software developments trying to avoid the mechanical part of the photogrammetric plotters, but more or less based on the same algorithms as analytical plotters. These machines, still devoted to professional practitioners, appeared a little cheaper, but with a comparatively poor visual quality. They were often considered as good workstations for orthophotographic productions with, as a side quality, an additional production capacity of photogrammetric plotting in case of emergency work. The main residual difficulty was linked to the huge amount of bytes, and also to the digitization of the traditional argentic aerial images, as the necessary scanners were extremely expensive. The high cost of the very precise mechanical parts had already been reduced in the 1980s from the old opto-mechanical plotters to the analytical plotters, as the mechanics has been downsized from 3D to 2D. But as a final revolution they had just shifted from the analytical plotter to the scanner, although a scanner could be used for many plotting workstations, and thus the total production cost was already lower. But this necessity to use a high-cost scanner was the main limitation of any new development of photogrammetric activity.

During this period, the industry developed for the consumer public very interesting products devoted to following the explosion of the microinformatics evolution and providing a set of extremely cheap tools: video cards allowing a fast and nearly real-time image management, stereoscopic glasses for video games, A4 scanners, digital colour cameras, etc. These developments were quickly taken into account by photogrammetrists, so that the Digital Photogrammetric Workstations (DPW) became cheaper and cheaper, most of the cost today being the software cost. This has completely changed the situation of photogrammetry: it is now possible to own a DPW and to use it only occasionally, as the acquisition costs are low. And if the user lacks a good stereoscopic vision capability, the automatic matching will help him to put the floating point in good contact with the surveyed zone. Of course, only trained people may quickly survey a large amount of data in a cost-effective way, but a good possibility to work is opened to non-specialists.

If the situation changed a lot for non-photogrammetrists, in fact it changed a lot for the photogrammetrists themselves, too: the main revolution we have met in the 1990s is the availability of aerial digital cameras, providing images whose quality is far superior to the classic argentic ones, at least in terms of radiometric fidelity, noise level, and geometric perfection. Thus the last obstacle for a wide spreading of digital photogrammetry is disappearing. We are on the verge of a massive transfer of such techniques toward purely digital processes, and the improvement in quality of the products, hopefully, should be very significant in the next few years. Also, the use of digital images allows us to automate more and more tasks, even if a large part of the automation is still at a research level (most automatic cue extractions).

We have then a completely new situation, where (for a basic activity at least) the equipment is no longer a limit for a wide spread of photogrammetric methodology. But if the use is now spreading rapidly, the theoretical aspects are not sufficiently known today, which could result in disappointing results for newcomers.

Thus this book has been oriented towards engineers and technicians, including technicians who are basically external to the field of photogrammetry and who would like to understand at least part of this field, for example to use it from time to time, without the help of professionals of photogrammetry. The goal here is to get this specialty much closer to the end user, and to help him to be autonomous in most circumstances, a little like word processors allowed us to type what we wanted when we wanted it, twenty years ago. But this end user must already have a scientific and technical culture, as the theoretical concepts would not be accessible without such a background. So I have supposed in this publication that the reader had such an initial understanding. Of course, professors and students, young ones as well as former specialists going through Continuous Professional Development, will also find great benefit from this up-to-date material.

A final very important point: the goal here being to be extremely close to the practical applications and to the production of data, we have looked for co-authors who are, each of them, excellent specialists in given aspects of digital photogrammetry, and most of them are practitioners or directly working for them. This is a guarantee that the considerations discussed here are really close to the application world. But the unavoidable drawback is that we have here 21 different authors for a single book, and that despite our efforts it is obvious that the homogeneity is not perfect. Some aspects are presented twice, sometimes with significant differences of point of view: this will remind each of us that the reality is complex, even in technical matters, and that controversy is the essence of our societies. In any case we have included an abundant bibliography that may be useful for further reading. Of course, as many authors are from France, most of the illustrations come from studies performed in this country, especially as IGN-F has been one of the few pioneers of the digital CCD cameras. But I think that this material is more or less equivalent to what everybody may find in his own country, especially now that large companies commercially propose digital cameras.

Michel Kasser

1 Image acquisition. Physical aspects. Instruments

INTRODUCTION

We may define photogrammetry as 'any measuring technique allowing the modelling of a 3D space using 2D images'. Of course, this is perfectly suitable for the case where one uses photographic pictures, but this is still the case when any other type of 2D acquisition device is used, for example a radar, or a scanning device: the photogrammetric process is basically independent of the image type. For that reason, we will present in this chapter the general physical aspects of the data acquisition in digital photogrammetry. It starts with §1.1, a presentation of the geometric aspects of photogrammetry, which is a classic problem but whose presentation is slightly different from books dealing with traditional photogrammetry. In §1.1 there are also some considerations about atmospheric refraction, the distortion due to the optics, and a few rules of thumb in planning aerial missions where digital images are used. In §1.2, we present some radiometric impacts of the atmosphere and of the optics on the image, which helps to understand the very physical nature of the aerial images. And in §1.3, these few lines about colorimetry are necessary to understand why the digital tools for managing black and white panchromatic images need not be the same as the tools for managing colour images. Then we will present the instruments used to obtain the images, digital of course. In §1.4 we will consider the geometric constraints on optical image acquisitions (aerial, and satellite as well). From §1.5 to §1.8, the main digital acquisition devices will be presented, CCD cameras (§1.5), radars (§1.6), airborne laser scanners (§1.7), and photographic image scanners (§1.8). And we will close this chapter with an analysis of the key problem of the size of the pixel, and the signal-to-noise ratio in the image.

1.1 MATHEMATICAL MODEL OF THE GEOMETRY OF THE AERIAL IMAGE

Yves Egels

The geometry of images is the basis of all photogrammetric processes, analogical, analytic or digital. Of course, the detailed geometry of an image essentially depends on the physical features of the sensor used. In analytic photogrammetry (the only difference with digital photogrammetry is the digital nature of the image) the use of this technology of image acquisition requires therefore a complete survey of its geometric features, and of the mathematical tools expressing the relation between the coordinates on the image and those of ground points (and, if the software is correct, this will be the only thing to do).

We shall explain here, just as an example, the case of the traditional conical photography (the image being silver halide or digital), that is still today the most used and that will be able to be a guide for the analysis of a different sensor. The leading principle, here as well as in any other geometrical situation, is to approach the quite complex real geometry by a simple mathematical formula (here the perspective) and to consider the differences between the physical reality and this formula as corrections ('corrections of systematic errors') presented independently. The imaging devices whose geometry is different are presented later in this chapter (§§1.4 to 1.7).

1.1.1 Mathematical perspective of a point of the space

The perspective of a point M of the space, of centre S, on the plane P is the intersection m of the line SM with the plane P. Coordinates of M will be expressed in the reference system (O, X, Y, Z), and the ones of m in the reference system of the image (c, x, y, z). The rotation matrix $\mathbf{R}$ corresponds to the change of system $(O, X, Y, Z) \to (c, x, y, z)$. One will note F the coordinates of the summit S in the reference of the image.

$$M = \begin{pmatrix} X_M \\ Y_M \\ Z_M \end{pmatrix} \quad S = \begin{pmatrix} X_S \\ Y_S \\ Z_S \end{pmatrix} \quad F = \begin{pmatrix} x_c \\ y_c \\ p \end{pmatrix} \quad m = \begin{pmatrix} x \\ y \\ 0 \end{pmatrix}$$

m image of $M \Leftrightarrow \vec{Fm} = \lambda.\vec{SM}$.

If one calls K the vector unit orthogonal to the plane of the image one can write:

$$m \in picture \Leftrightarrow K^t m = 0 \text{ that implies } \lambda = \frac{-K^t F}{K^t R(M - S)},$$

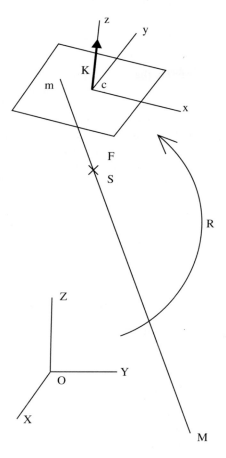

Figure 1.1.1 Systems of coordinates for the photogrammetry.

from where the basic equation of the photogrammetry, so-called collinearity equation:

$$m = F - \frac{K^t FR(M-S)}{K^t R(M-S)}. \tag{1.1}$$

This equation can be developed and reordered according to coordinates of M, under the so-called '11 parameters equation', which appears simpler to use, but should be handled with great care:

$$x = \frac{a_1 X_M + b_1 Y_M + c_1 Z_M + d_1}{a_3 X_M + b_3 Y_M + c_3 Z_M + 1} \quad y = \frac{a_2 X_M + b_2 Y_M + c_2 Z_M + d_2}{a_3 X_M + b_3 Y_M + c_3 Z_M + 1}. \tag{1.2}$$

1.1.2 Some mathematical tools

1.1.2.1 Representation of the rotation matrixes, exponential and axiator

In computations of analytic photogrammetry, the rotation matrixes appear quite often. On a vector space of n dimensions, they depend on $n(n-1)/2$ parameters, but do not form a vector sub-space. There is therefore no linear formula between a rotation matrix and its 'components'. It is, however, vital to be able to express such a matrix judiciously according to the chosen parameters. In the case of R^3, it is usual to analyse the rotation as the composition of three rotations around the three axes of coordinates (angles w, ϕ and κ, familiar to photogrammetrists). But the resulting formulas are laborious enough and lack symmetry, which obliges us to perform unnecessary developments of formulas. One will be able to define a rotation of the following manner physically: rotation of an angle θ around a unitary vector

$$\vec{\Omega} = \begin{pmatrix} a \\ b \\ c \end{pmatrix}, \text{ with } (\sqrt{a^2 + b^2 + c^2} = 1)$$

We will call here the axiator of a vector the matrix equivalent to a vectorial product:

$$\vec{\Omega} \wedge \vec{V} = \tilde{\Omega} \cdot V$$

one will check that

$$\tilde{\Omega} = \begin{pmatrix} 0 & -c & b \\ c & 0 & -a \\ -b & a & 0 \end{pmatrix}. \tag{1.3}$$

It is easy to control that the previous rotation can be written: $R = e^{\tilde{\Omega}\theta}$. Indeed,

$$e^{\tilde{\Omega}\theta} = I + \tilde{\Omega}\theta + \ldots + \frac{\tilde{\Omega}^n \theta^n}{n!} + \ldots, \tag{1.4}$$

therefore, since $\tilde{\Omega} \cdot \Omega = \vec{\Omega} \wedge \vec{\Omega} = 0$ (a well-known property of the vectorial product) $e^{\tilde{\Omega}\theta} \cdot \Omega = \vec{\Omega}$. $\vec{\Omega}$ is the rotation axis, only invariant. One will check that $\tilde{\Omega}^3 = -\Omega$, $\tilde{\Omega}^4 = -\tilde{\Omega}^2$, etc.; on the other hand, $\tilde{\Omega}^t = -\tilde{\Omega}$

$$(e^{\tilde{\Omega}\theta}) = I + \tilde{\Omega}^t\theta + \ldots + \frac{\tilde{\Omega}^{t^n}\theta^n}{n!} + \ldots$$

$$= I + \tilde{\Omega}\theta + \ldots + (-1)\frac{\tilde{\Omega}^n \theta^n}{n!} + \ldots$$

$$= e^{-\tilde{\Omega}\theta} = (e^{\tilde{\Omega}\theta})^{-1}. \qquad (1.5)$$

Therefore this matrix is certainly an orthogonal matrix, which is a rotation matrix as det $\mathbf{R} = 1$ (as $\tilde{\Omega}^t = -\tilde{\Omega}$). One will also verify that the rotation angle is equal to θ, for example by computing the scalar product of a normal vector to Ω with its transform.

One sees therefore, while regrouping the even terms and the odd terms of the procedure:

$$R = I + \tilde{\Omega}\theta - \frac{\tilde{\Omega}\theta^3}{3!} + \ldots + \frac{\tilde{\Omega}^2 \theta^2}{2!} - \frac{\tilde{\Omega}^2 \theta^4}{4!} + \ldots$$

$$R = I + \tilde{\Omega}\sin\theta + \tilde{\Omega}^2(1 - \cos\theta) \quad \text{(Euler formula)}. \qquad (1.6)$$

This formula will permit a very comfortable calculation of rotations. Indeed one can choose, as parameters of the rotation, the three values $a\theta$, $b\theta$ and $c\theta$. One can then write:

$$\vec{\Theta} = \vec{\Omega}\theta = \begin{pmatrix} a\theta \\ b\theta \\ c\theta \end{pmatrix} \text{ and thus } \theta = \|\vec{\Theta}\|, \vec{\Omega} = \frac{\vec{\Theta}}{\theta}. \qquad (1.7)$$

If one expresses $\sin\theta$ and $\cos\theta$ using $\tan\theta/2$ with the help of the classic trigonometric formulas, it becomes:

$$R = (I - \tilde{\Omega}\tan(\theta/2))^{-1}(I + \tilde{\Omega}\tan(\theta/2)) \quad \text{(Thomson formula)}. \qquad (1.8)$$

1.1.2.2 Differential of a rotation matrix

Expressed under this exponential shape, it is possible to differentiate the rotation matrixes. One may determine:

$$dR \approx e^{\tilde{\Theta}} d\tilde{\Theta} = R \, d\tilde{\Theta}. \qquad (1.9)$$

In this expression, the differential makes the parameters of the rotation appear under matrix shape, which is not very practical. Applied to a vector, it can be written (using the anticommutativity of the vectorial product):

$$dR \cdot A = R \cdot d\tilde{\Theta} \cdot A = -R \cdot \tilde{A} \cdot d\Theta. \qquad (1.10)$$

1.1.2.3 Differential relation binding m to the parameters of the perspective

The previous equation is not linear, and to solve systems corresponding to analytical photogrammetry (orientation of images, aerotriangulation), it is necessary to linearize them. Variables that will be taken into account are F, M, S and R.

If one puts $A = M - S$ and $U = RA$

$$dm = dF - \frac{K^t dF \, U}{K^t U} - \left(\frac{K^t F}{K^t U} - \frac{K^t F U K^t}{(K^t U)^2} \right) dU \tag{1.11}$$

$$dU = R(dM - dS) + dRA = R(dM - dS - \tilde{A} \, d\Theta). \tag{1.12}$$

On the other hand, $K^t F$ being a scalar, $K^t F U = U K^t F$ and $K^t dF \, U = U K^t \, dF$.

One then gets:

$$dm = \frac{K^t U - U K^t}{(K^t U)^2} [K^t U \, dF - K^t F R(dM - dS - \tilde{A} \, d\Theta)]. \tag{1.13}$$

If one writes

$$p = K^t F, \quad U = \begin{pmatrix} u_1 \\ u_2 \\ u_3 \end{pmatrix} \quad \text{and} \quad V = \begin{pmatrix} u_3 & 0 & -u_1 \\ 0 & u_3 & -u_2 \end{pmatrix},$$

one gets after manipulation:

$$\begin{pmatrix} dx \\ dy \end{pmatrix} = \frac{V}{u_3} dF + \frac{p}{u_3^2} VR(dS - dM) + \frac{p}{u_3^2} VR \tilde{A} \, d\Theta. \tag{1.14}$$

Under this matrix shape, the computer implementation of this equation requires only a few code lines.

1.1.3 True geometry of images

In physical reality, the geometry of images only reproduces in an approximate way the mathematical perspective whose formulas have just been established. If one follows the light ray from its starting point on the ground until the formation of the image on the sensor, one will find a certain number of differences between the perspective model and the reality. It is difficult to be comprehensive in this area, and one will mention here only the main reasons for distortion whose consequences are appreciable in the photogrammetric process.

1.1.3.1 Correction of Earth curvature

The first reason for which photogrammetrists use the so-called correction of Earth curvature in photogrammetry is that the space in which cartographers operate is not the Euclidian space in which the collinearity equation is set up. It is indeed in a cartographic system, in which the planimetry is measured in a plane representation (often conform) of the ellipsoid, and the altimetry in a system of altitudes.

The best solution for dealing with this is to make rigorous transformations from cartographic to tridimensional frames (it is usually more convenient to choose a local tangent tridimensional reference, if it is necessary just to write relations of planimetric or altimetric conversion in a simple way). This solution, if better in theory, is, however, a little difficult to put into an industrial operation, because it requires having a database of geodesic systems, of projection algorithms (sometimes bizarre) used in the client countries, and arranging a model of the zero level surface (close to the geoid) of the altimetric system concerned.

An approximate solution to this problem can nevertheless be found, that is – experience proves it – of comparable precision with the precision of this measurement technique. It relies on the fact that all conform projections are locally equivalent, with only a changing scale factor. One can attempt to replace the real projection, in which is expressed the coordinates of the ground, by a unique conform projection therefore (and why not try for simplicity!), for example a stereographic oblique projection of the mean curvature sphere on the tangent plane to the centre of the work.

The mathematical formulation then becomes as shown in Figure 1.1.2 (the figure is made in the diametrical plane of the sphere containing the centre of the work A and the point I to transform).

I is the point to transform (cartographic), J the result (tridimensional), I' the projection of I on the plane and J' the projection of J on the terrestrial sphere of R radius. I' and J' are the inverse of each other in an inversion of B pole and coefficient $4R^2$.

Let us put:

$$I = \begin{Bmatrix} x \\ R+h \end{Bmatrix} \quad I' = \begin{Bmatrix} x \\ R \end{Bmatrix} \quad J' = \begin{Bmatrix} x' \\ y' \end{Bmatrix} \quad J = (1+\beta)\begin{Bmatrix} x' \\ y' \end{Bmatrix} \quad \alpha = \frac{x}{2R} \quad \beta = \frac{h}{R}.$$

The calculation of the inversion gives:

$$\left.\begin{matrix} \|BI'\| \cdot \|BJ'\| = 4R^2 \\ BJ' = \lambda BI' \end{matrix}\right\} \Rightarrow \lambda = \frac{4R^2}{BI'^2} = \frac{1}{1+\alpha^2} \approx 1 - \alpha^2 \qquad (1.15)$$

$$J' = B + \lambda B' = \begin{Bmatrix} x(1-\alpha^2) \\ R(1-2\alpha^2) \end{Bmatrix} \quad J = \begin{Bmatrix} (1-\alpha^2)(1+\beta)x \\ (1-2\alpha^2)(1+\beta)R \end{Bmatrix}$$

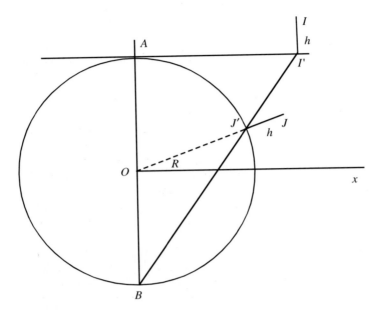

Figure 1.1.2 Figure made in the diametrical plane of the sphere containing the centre of the work A and the point I to transform.

$$J - I = \begin{cases} (\beta - \alpha^2)x \\ -2\alpha^2 R \end{cases}. \qquad (1.16)$$

One can consider that $J - I$ corresponds to a correction to bring us to the coordinates of I:

$$\Delta I_{\text{curvature}} = \begin{cases} xh/R - x^3/4R^2 \\ -x^2/2R \end{cases}. \qquad (1.17)$$

In the current conditions, the term of altimetric correction is by far the most meaningful. But for aerotriangulations of large extent, planimetric corrections can become important, and must be considered. If the aerotriangulation relies on measures of an airborne GPS (global positioning system), it is necessary not to forget also to do this correction on coordinates of tops, that also have to be corrected for the geoid–ellipsoid discrepancy.

1.1.3.2 Atmospheric refraction

Before penetrating the photographic camera, the luminous rays cross the atmosphere, whose density, and therefore refractive index, decreases with the altitude (see Table 1.1.3).

Table 1.1.3 Atmospheric co-index of refraction–variations with the altitude

H (km)	−1	0	1	3	5	7	9	11
$N = (n-1)\cdot 10^6$	306	278	252	206	167	134	106	83

This variation of the index provokes a curvature of the luminous rays (oriented downwards) whose amplitude depends on atmospheric conditions (pressure and temperature at the time of the image acquisition), that are not uniform in the field of the camera, and are generally unknown.

Nevertheless, this deviation is small with regard to the precision of photogrammetric measurements (with the exception of very small scales). It will thus be acceptable to correct the refraction using a standard reference atmosphere: in order to evaluate the influence of the refraction, it will be necessary to use a model providing the refraction index at various altitudes.

Numerous formulas exist allowing an evaluation of the variation of the air density according to altitude. No one is certainly better that any other, because of the strong turbulence of low atmospheric layers. In the tropospheric domain (these formulas would not be valid if the sensor is situated in the stratosphere), one will be able to use, for example, the formula published by the OACI (Organisation de l'Aviation Civile Internationale):

$$n = n_0 (1 + ah + bh^2), \qquad (1.18)$$

where

$a = -2.560.10^{-8}, b = 7.5.10^{-13}, n_0 = 1.000278$ h in metres,

or this one, whose integration is simpler:

$$n = 1 + ae^{-bh}, \qquad (1.19)$$

where

$a = 278.10^{-6}, b = 105.10^{-6}$.

The rays appear to come then from a point situated in the extension of the tangent to the ray at the entrance into the optics, thus introducing a displacement of the image.

The calculation of the refraction correction can be performed in the following way: the luminous ray coming from M seems to originate from the point M' situated on one same vertical, so that $MM' = y$. One will modify the altitude of the object point M by the correction y by putting it in M'. This method of calculation has the advantage of not supposing

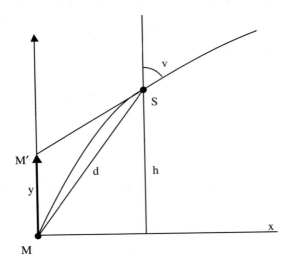

Figure 1.1.4 Optical ray in the atmosphere (with a strong exaggeration of the curvature).

that the axis of the acquired image is vertical, and thus works also in oblique terrestrial photogrammetry. (See Figure 1.1.4.)

$$dy = \frac{h - h_M}{\sin v \cdot \cos v} dv,$$

from where

$$y = MM' = \int_M^S \frac{h - h_M}{\sin v \cdot \cos v}. \tag{1.20}$$

The formula of Descartes gives us the variation of the refraction angle:

$$n \cdot \sin v = cte,$$

or $\quad dn \sin v + n \cos v \, dv = 0,$

or again

$$dv = -\frac{dn}{n} \tan v. \tag{1.21}$$

In the case of photography of a zone of small extension, one will suppose the Earth to be flat, the vertical (and the refractive index gradients) are therefore all parallel to each other (this approximation is not valid in spatial imagery). The radius being very large, one will be able to assimilate the bow and the chord, the v angle and the refractive index can be considered constant for the integration.

$$MM' = -\int_M^S \frac{(h - h_M)}{\sin v \cdot \cos v} \frac{\sin v}{\cos v} \frac{dn}{n} = -\int_M^S \frac{(h - h_M)}{\cos^2 v} \frac{dn}{n}$$

$$= \frac{-1}{\cos^2 v} \int_M^S (h - h_M) \frac{dn}{dh} dh \ . \tag{1.22}$$

With this second formula for refraction, the calculation for the complete rays gives:

$$\frac{dn}{dh} = -ab \, e^{-bh}, \quad \int \frac{dn}{dh} dh = a \, e^{-bh} = n - 1$$

and

$$\int h \frac{dn}{dh} dh = \left(h + \frac{1}{b}\right)(n - 1) \ . \tag{1.23}$$

If one puts $D = \|SM\|$ and $H = h_S - h_M$ one gets the following refraction correction:

$$y = -\frac{D^2}{H^2} \left[\frac{n_S - n_M}{b} + H(n_S - 1) \right] . \tag{1.24}$$

For aerial pictures at the 1/50,000th scale with a focal length of 152 mm, this correction is of about 0.20 m to the centre, and reaches 0.50 m at the edge of the image. It is appreciably less important at larger scales for the same focal length.

This formula also permits either the correction of refraction with oblique or horizontal optical axes lines (terrestrial photogrammetry), which are very obviously not acceptable domains in very commonly used formulas, bringing a radial correction to the image coordinates. One will be able to verify that when $h_M \to h_S$, the limit of y is:

$$y_H = (n_S - 1) \frac{b}{2} D^2,$$

the classical refraction formula for a horizontal optical axis.

1.1.3.3 At the entrance in the plane

In a pressurized plane two phenomena may occur. First, the light ray crosses the glass plate placed before the objective. This glass plate, therefore of weak thermal conductivity, is subject to significant temperature differences and bends under the differential dilation effect (in the opposite direction to the bending due to the pressure difference, that induces a much smaller effect). It takes a more or less spherical shape, and operates then like a lens that introduces an optic distortion. Unfortunately, if the phenomenon can be experimentally tested (and the calculation shows that its influence is not negligible), it is practically impossible to quantify it, the real shape of the glass plate porthole depends on the history of the flight (temperatures, pressures inside and outside). There are techniques of calculation that will be applied to §1.1.3.6, however, (a posteriori unknowns) permitting one to model and to eliminate the influence of it in the aerotriangulation.

When the plane is pressurized the luminous ray is refracted a second time at the entrance to the cabin, where the pressure is higher than that outside at the altitude of flight (the difference in pressure is a function of the airplane used). One can model this refraction in the same way as for the atmospheric refraction (the sign is opposed to the previous one, because the radius arrives in a superior index medium).

While noting n_c the refractive index inside the cabin, and while supposing that the porthole is horizontal,

$$dv = -(n_c - n_s) \tan v = -(n_c - n_s) \frac{R}{H} \tag{1.25}$$

$$dy = -\frac{D^2}{H^2} (n_c - n_s) H. \tag{1.26}$$

While combining this equation with that of atmospheric refraction, one sees that it is sufficient to replace the altitude of the top by the altitude of the cabin in the second part:

$$y = -\frac{D^2}{H^2} \left[\frac{n_S - n_M}{b} + H(n_c - 1) \right]. \tag{1.27}$$

Thus we have to compute the cabin index of refraction. Often in pressurized planes, the pressure is limited to a maximal value of ΔP_{max} (for example 6 psi = 41,370 Pa for the Beechcraft King Air of IGN-F) between inside and outside. Under a given altitude limit (around 4,300 m in the previous example), the pressure is kept to the ground value. The refractive index n and co-index N are linked to the temperature and pressure by an approximate law:

$$N = (n - 1)10^6 = 790 \frac{P}{T}, \quad P \text{ in kPa and } T \text{ in K.}$$

The temperature decreases with the altitude with an empirical variation $T = T_0 - 6.5 \times 10^{-3} Z$. If one takes $T_C = T_0 = 298$ K, the cabin refractive index will be:

$$N_C = N_S (1 - 2.22 \times 10^{-5} Z_S) + 2{,}696 \, \Delta P.$$

This index will be limited to its ground value ($N = 278$ in this model), if the altitude is below the pressurization limit. One may note that, in current flight conditions (flight altitude 5,000 m, pressurization limit 4,300 m), this total correction is nearly the opposite of the correction computed without cabin pressurization.

1.1.3.4 Distortion of the optics

In elementary optics (conditions of Gauss: thin lenses, rays with a low tilt angle with respect to the optical axis, low aperture optics), all rays passing through the centre of the objective are not deviated. This replicates the mathematical perspective. Unfortunately, the real optics used in photogrammetric image acquisitions do not fulfil these conditions.

In the case of perfect optics (centred spherical dioptres) the entrance ray, the exit ray, and the optical axis are coplanar. But the incident and emergent rays are not parallel. This gap of parallelism constitutes the distortion of the objective. It is a permanent and steady characteristic of the optics, which can be measured prior to the image acquisitions, and corrected at the time of photogrammetric calculations.

The measure of this distortion can be achieved on an optical bench, using collimators, or by observation of a dense network of targets through the objective. Manufacturers, to provide certification of the aerial camera that they produce, use these methods, which are quite complicated and costly. It is also possible to determine the distortion (as well as the centring and the principal distance) by taking images of a tridimensional polygon of points very well known in coordinates, and by calculation of the collinearity equations, to which one adds unknowns of distortion according to the chosen model.

In the case of perfect optics, the distortion is very correctly modelled by a radial symmetrical correction polynomial around the main point of symmetry (intersection of the optic axis with the plane of the sensor). Besides, one can demonstrate easily that this polynomial only contains some terms of odd rank. The term of first degree can be chosen arbitrarily, because it is bound by choice of the main distance: one often takes it as a null (main distance to the centre of the image) or so that the maximal distortion in the field is the weakest possible.

If the optics cannot be considered as perfect (it is sometimes the case if one attempts to use cameras designed for the general public for photogrammetric operations), the distortion can lose its radial and symmetrical character, and requires a means of calibration and more complex modelling. Besides, in the case of zooms (that one should avoid ...), this distortion does not remain constant in time because of variations of centrage brought about by the displacement of the pancratic vehicle (mobile part permitting the displacement of lenses). Any modelling then becomes very uncertain.

1.1.3.5 Distortions of the sensor

The equation of the perspective supposes that the sensor is a plane, and that one can define there a reference of fixed measure. Thus, it will be necessary to compare the photochemical sensors (photographic emulsion) and the electronic sensors (CCD matrixes). These last are practically indeformable, and the image that they acquire is definitely steady (in any case on the geometric plane). The following remarks will therefore apply to the photochemical sensors only.

The planarity of the emulsion was formerly obtained by the use of photographic glass plates, but this solution is no longer used, except maybe in certain systems of terrestrial photogrammetry. It was, besides, not very satisfactory. Today one practically always uses vacuum depression of the film on the plate at the bottom of the camera. This method is very convenient, provided that there is no dust interposed between the film and the plate, which is unfortunately quite difficult to achieve. There follows a local distortion of the image, which is difficult to detect (except in radiometrically uniform zones, where there appears an effect of shade) and impossible to model.

The dimensional stability of the image during the time, if it is improved with the use of so-called 'stable' supports, is quite insufficient for the use of photogrammetrists. Distortions ten times better than the necessary measurement precision are frequent. The only efficient counter-measure is the measure of the coordinates on the images of well-known points (reference marks of the bottom plate of the camera) and the modelling of the distortion by a bidimensional transformation. One usually chooses an affinity (general linear transformation) that represents correctly the main part of the distortion. But this distortion is thus measured only on the side of the image, in a limited number of points (eight in general). The interpolation inside the image is therefore of low reliability.

1.1.3.6 Modelling of non-quantifiable defects

The main corrections to apply to the simplistic model of the central perspective have been reviewed. Several of them can be considered as perfectly known – the curvature of the Earth for example; others can be calculated roughly: refraction, distortion of the film, and others that are completely unknown, the distortion of the porthole glass plate, for example.

When the precise geometric reconstitution is indispensable – as is notably the case in aerotriangulation, where the addition of systematic errors can generate intolerable imprecision – a counter-measure consists in adding to the equation of collinearity and its corrective terms a parametric model of distortion (often polynomial). This model contains a small number of parameters, chosen in order to best represent influences of no directly modelizable defects. These supplementary a posteriori unknowns of systematism will be solved simultaneously with the main unknowns of the photogrammetric system, which are coordinates of the ground points, tops of images, and rotations of images. This technique usually permits a gain of 30 to 40 per cent on altimetric precision of the aerotriangulation.

1.1.4 Some convenient rules

Rules for the correct use of aerial photogrammetry have been known for several decades, and we will therefore only briefly recall the main features. Besides, variables on which one can intervene are not numerous. This arises essentially from the necessary precision, which can be specified in the requirements. They can also be deduced from the scale of the survey (even though this one is digital, one can assign it a scale corresponding to its precision, that is conventionally 0.1 mm to the scale). The practice shows that the errors that accumulate during the photogrammetric process limit its precision – in the current studies – to about 15 to 20 microns on the image (on the very well-defined points) for each of the two planimetric coordinates. This number can be improved appreciably (down to 2 to 3 μm) by choosing some more coercive operative (and simultaneously costlier) ways to work. For the altimetry, this number must be grosso modo divided by the ratio between the basis (distance between points of view) and the distance (of these points of view to the photographed object). This ratio depends directly on the chosen field angle for the camera: the wider the field, the more this ratio will increase, the more the altimetric precision will be close to the planimetric precision. As an example, Table 1.1.5 sums up the characteristics of the most current aerial camera values (in format 24×24 cm) with values of standard base length (overlap of 55 per cent).

Knowing the precision requested for the survey, it is then very simple to determine the scale of the images. The choice of the focal distance will be guided by the following considerations: the longer the focal distance, the higher the altitude of flight will be, the more the hidden parts will be reduced (essentially feet of buildings in city, not to mention complete streets), but simultaneously the more the altimetric precision is degraded, and the more marked the radiometric influence of the atmosphere.

Table 1.1.5 Characteristics of the most current aerial camera values (in format 24 × 24 cm) with values of standard baselength (overlap of 55 per cent)

Focal length	88 mm	152 mm	210 mm	300 mm
Field of view	120°	90°	73°	55°
B/H	1.3	0.66	0.48	0.33
Flight height (1/10,000)	880 m	1520 m	2100 m	3000 m
Planimetric precision (1/10,000)	20 cm	20 cm	20 cm	20 cm
Altimetric precision (1/10,000)	15 cm	30 cm	40 cm	60 cm

In practice, one will preferentially use the focal distance of 152 mm. Shorter focal lengths will be reserved for the very small scales, for which the altitude of flight can constitute a limitation (the gain in altimetric precision is actually quite theoretical, as the mediocre quality of the optics makes one lose what was gained in geometry). A long focal distance will be preferred most of the time for the orthophotographic surveys in urban zones, in which the loss of altimetric quality is not a major handicap. In the case of a colour photograph, it will be necessary nevertheless to take care of the increase of the atmospheric diffusion fog (due to a thicker diffusing layer), which will especially have the consequence of limiting the number of days when it is possible to take the images. Another way (less economic) to palliate the hidden part problem is to use a camera with a normal focal distance, while choosing more important overlaps: one can reconcile a good altimetric precision, a weak atmospheric fog, and acceptable distortions for objects on the ground.

1.2 RADIOMETRIC EFFECTS OF THE ATMOSPHERE AND THE OPTICS

Michel Kasser

1.2.1 Atmospheric diffusion: Rayleigh and Mie diffusions

The atmosphere is generally opaque to the very short wavelengths, its transmission only starting toward 0.35 μm. Then, the atmosphere presents several 'windows' until 14 μm, the absorption becoming again practically total between 14 μm and 1 mm. Finally, the transmission reappears progressively between 1 mm and 5 cm, to become practically perfect to the longer wavelengths (radio waves).

The low atmosphere, in the visible band, is not perfectly transparent, and when it is illuminated by the Sun it distributes a certain part of the radiance that it receives: all of us can observe the milky white colour of the sky close to the horizon, and to the contrary its blue colour is very marked in high altitude sites towards the zenith. These luminous emissions are important data to understand correctly the images used in photogrammetry, and in particular the diffuse lighting coming from the sky in zones of shades, or again the atmospheric fog or white-out and its variation according to the altitude of image acquisition. We will start with a presentation of the present diffusion mechanisms in the atmosphere, to assess the effects of this diffusion on the pictures obtained from an airplane or a satellite.

1.2.1.1 Rayleigh diffusion

The diffusion by the atmospheric gases (often called absorption, in an erroneous way, since the luminous energy removed from the light rays is merely redistributed in all directions, but not transformed into heat) is due to the electronic transitions or vibrations of atoms and molecules generally present in the atmosphere. On the whole, the electronic transitions of atoms provoke a diffusion generally situated in the UV (these resonances are very much damped because of the important couplings between atoms in every molecule, and thus the effect is still noticeable even at much lower frequencies), whereas vibration transitions of molecules rather provoke diffusions situated at lower frequencies, and therefore in the IR (and these are sharp resonances, because there is nearly no coupling between the elementary resonators that are molecules, and therefore the flanks of bands are very stiff). The Rayleigh diffusion, particularly important to the high optical frequencies, corresponds to these atomic resonances. We have therefore a factor of attenuation of a light ray when it crosses the atmosphere, and at the same time a substantial source of parasitic light, since light removed from the incidental ray is redistributed in all directions, so that the atmosphere becomes a secondary light source.

In short, Rayleigh diffusion is characterized by two aspects:

- its efficiency varies as λ^{-4}, which combined with the spectral sensitivity of the human eye is responsible of the blue appearance of the sky;
- the diagram of light redistributed is symmetrical between the before and the rear, with a front or rear diffusion twice as important as the lateral one.

This diffusion may thus be very satisfactorily and accurately modelled.

1.2.1.2 Diffusion by sprays (Mie diffusion)

One calls sprays all particles, droplets of water, dusts that are continuously sent into the atmosphere by wind sweeping soil, but that can come also

Table 1.2.1 Typical particle size in various sprays

Smoke	0.001–0.5 μm
Industrial smoke	0.5–50 μm
Mist, dust	0.001–0.5 μm
Fog, cloud	2–30 μm
Drizzle	50–200 μm
Rain	200–2000 μm

from the sea (particles of salt), from volcanoes because of cataclysmic explosions, from human activities (industrial pollution, cars, forest fires, etc.) and also from outside of our globe: meteoroids and their remnants. The essential characteristic of these sprays, since such is the generic name that one usually gives to all of these non-gaseous components of the atmosphere, is that their distribution is at any instance very variable with the altitude (presenting a very accentuated maximum to the neighbourhood of the ground) but also very variable with the site and with time. (See Table 1.2.1.)

In fact, every spray distributes light, in the same way that gaseous molecules do. However, an important difference is that molecules present some weaker dimensions than the wavelength of light that interests us here. This implies that the Rayleigh diffusion by molecules (or atoms) is more or less isotropic. To the contrary, the diffusion by sprays, whose dimensions are often the same order as the wavelengths considered, present strong variations with the direction of observation. At the same time, the intensity of the diffusion by sprays – also called the Mie diffusion, depends strongly on the dimension of the diffusing particle. The exact calculation is of limited value, since in practice the quantity of sprays in the atmosphere is very variable according to the altitude, and also especially according to the time. In any case, it is generally not possible to foresee it.

In practice, it is to the contrary from luminous diffusion measurements that one may measure the density (in number of particles/cm^3) of sprays according to the altitude.

Table 1.2.2 provides some figures, but it is necessary to note that this distribution is reasonably steady only to an altitude over 5 km. In the first kilometres of the atmosphere, to the contrary, the influence of wind, of human activities is such that numbers in the table are only indicative.

The Mie diffusion is characterized by a directional diagram that varies markedly with the size of sprays. For the very small particles compared to the wavelength, the diagram is very similar to that of Rayleigh diffusion. For larger particles, it becomes more and more dissymetric between front

Table 1.2.2 Typical spray density vs. altitude

h (km)	Sprays density (particles per cm³)
0	200
1	87
2	38
3	16
4	7.2
5	3.1
6	1.1
7	0.4
8	0.14
9	0.05

and rear, to give a retrodiffusion negligible compared to the importance of the front diffusion. (See Figure 1.2.3.)

To assess the importance of the phenomenon on aerial image acquisitions, one determines the attenuation of a light ray travelling upwards, from the observed horizontal attenuation, which is obviously more accessible. One characterizes it by the 'meteorological visibility' V, that is the distance to which the intensity of a collimated light ray of $\lambda = 0.555$ μm (wavelength of the maximum sensitivity the eye) falls to 2 per cent of its initial value. Under these conditions, the practice shows that the diffusion that comes with the attenuation removes all observable details of the observed zone, that melts in a uniform grey colour.

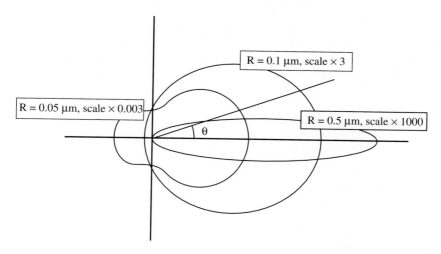

Figure 1.2.3 Angular distribution of the Mie diffusion for a wavelength of 0.5 μm and for three particle sizes: 0.05, 0.1 and 0.5 μm.

The reduction of intensity of a light ray depends, physically, on the number of particles interacting with it, the resulting attenuation being correctly described by Beer's law:

$$I = I_0 e^{-kx}, \qquad (1.28)$$

where x is the distance in the absorbing medium, and k the coefficient of absorption.

The visibility V, easily appreciated by a trained observer, allows the calculation $K_M + K_R$ and therefore the part of attenuation due to sprays, by the difference between the observed coefficient K_{obs} (such as: $e^{-K_{obs}V} = 0.02$) and the one K_R easy to calculate, resulting from Rayleigh diffusion.

$$I = I_0 e^{-(K_R+K_M)x} = I_0 e^{-K_R x} e^{-K_M x}. \qquad (1.29)$$

From the visibility V, one establishes in an experimental way the coefficient of attenuation for different wavelengths. This leads to the formula (in km^{-1}, for λ in µm and V in km):

$$K_M = \frac{3.91}{V} \left(\frac{0.555}{\lambda}\right)^{0.585 V^{1/3}}. \qquad (1.30)$$

Table 1.2.4 gives the numeric figures for various wavelengths.

If one wants to switch from the horizontal attenuation to the vertical attenuation, to assess the effect on a light ray going from the ground to a plane, one admits besides that the density of sprays, according to the altitude, follows an exponential law. Figure 1.2.5 shows the relation between the coefficient of extinction (that is the k coefficient of Beer's law) and the visibility.

Figure 1.2.6 illustrates the result of the integration calculation for a vertical crossing of the diffusing layer, done for an attenuation $K_M = 10^{-3}$ km^{-1} at $h = 5$ km (altitude over which the concentration of sprays stops being influenced considerably by ground effects, and may be considered as constant).

Table 1.2.4 Variations of the Mie and Rayleigh coefficients of attenuation at various wavelengths for three values of the ground meteorological visibility

λ(µm)	$K_{Rayleigh}$ (km^{-1})	K_{Mie} (km^{-1})		
		$V = 1$ km	$V = 5$ km	$V = 10$ km
0.4	0.044	4.7	1.1	0.58
0.6	0.0083	3.7	0.72	0.35
0.8	0.0026	3.1	0.54	0.24
1.0	0.0006	2.8	0.43	0.18

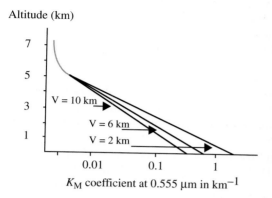

Figure 1.2.5 Variations of the coefficient of attenuation at 0.555 μm for three values of the ground meteorological visibility.

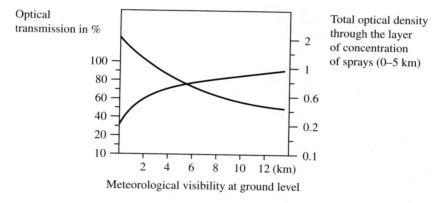

Figure 1.2.6 Optical transmission and optical thickness for a vertical crossing of the diffusing layers for various values of meteorological visibility at ground level.

One should also note that, in the visible domain, the content of water vapour does not change the visibility.

1.2.2 Effects of the atmospheric diffusion in aerial image acquisition

The diffusion of light by atmospheric gas molecules and by all sorts of sprays dispersed in the atmosphere is not only responsible for the light attenuation. In fact, the portion of atmosphere crossed by the light rays of observation behaves, when it is illuminated by the Sun, like a luminous source whose radiance arrives in part on the detector.

Being foreign to the ground that one generally intends to observe, this radiance can be qualified as 'parasitic', and constitutes one of the main sources of 'noise' in image processing in the visible domain. This diffusion is as difficult to calculate as the corresponding attenuation, as like it, it depends on distribution by sizes and by the nature of diffusing particles. And moreover it depends, in the case of sprays whose size is not negligible in comparison to λ, on the angle between the direction of observation and the direction of lighting. The contribution of the Rayleigh diffusion, especially important to the short wavelengths, is therefore the only one easy to calculate.

It remains to determine the fraction of this flux sent back towards the instrument of observation. To a first approximation, one will be able to admit that the diffusion is isotropic. The diffusion (sum of contributing diffusions Mie and Rayleigh) intervenes finally in the three following aspects:

- *Lighting brought by the sky in zones to the shade,* diffuse source that includes at least the Rayleigh diffusion, and that is therefore more important in the blue, but that is in any case characteristic of the shade contrast. If this contrast is strong, it means that shades are poorly illuminated, and therefore that the diffusion is weak. The lower the Sun is on the horizon, the weaker the direct solar lighting and the more the atmospheric diffusion becomes predominant. On an indicative basis, between a surface exposed to the Sun and the same surface in the shade, the typical lighting differences are of: 2 (Sun to 20° on the horizon, sky misty enough, $V = 2$ km), 3 (Sun to 45°, $V = 5$ km), 10 (extreme case, very high Sun and $V > 10$ km, frequent case, for example, in high mountains).
- *Attenuation of the useful light ray* going from the illuminated ground towards the device used for image acquisitions. This attenuation depends on the height of flight H and the value of V. For example for $H > 4$ km, the attenuation is the same as that of the direct solar radiance, and it may range from a value of transmission of 0.90 ($V = 10$ km, for example) to 0.70 ($V = 3$ km) or even less. And for the lower flight heights, the exponential decrease of K with the increasing altitudes shows that this coefficient is near unity for the low flights (large scales, $H = 1,500$ to $2,000$ m). On the other hand, for values of $H > 4$ or 5 km, the attenuation is the same as for an imaging satellite.
- *Superposition of a reasonably uniform atmospheric fog or white-out* on the useful picture formed in the focal plane. This parasitic luminous flux can be computed according to the following way in the case where the plane is entirely over the diffusing layer ($H > 4$ or 5 km): (1) the diffusion is reasonably symmetrical fore/rear; (2) this diffusion essentially originates from the light removed from the direct solar rays, between 10 per cent and 50 per cent according to our previous remarks;

(3) the half of this flux going upwards therefore represents the source of light, coming from all the lower half-space, the other half going downwards; (4) one makes the hypothesis that this flux is isotropic in all this half-space and one calculates, considering the aperture of the camera, the flux F that enters there from all this half-space; (5) one calculates the solid angle under which is seen a pixel on the ground by the camera; (6) one finally corrects F of the ratio formed by this solid angle and 2π (solid angle formed by a half-space). This figure for the flux will be reduced if the plane flies lower, as mentioned previously, but one is generally surprised when one does this calculation, as the atmospheric fog is so important as compared to the useful signal. In current conditions, for $H > 5$ km the energy from the atmospheric fog is frequently more than the energy of the useful image, which almost always gives to the uncorrected colour pictures a milky white aspect that makes the appearance of colours often disappear to a large extent.

1.2.3 Radiometric problems due to the optics

In the optic energy calculations that one does in aerial imagery, it is necessary to also take into account the classic effects due to camera optics. We have been used, with traditional aerial cameras, to seeing these defects corrected in a quasi-perfect way, essentially because the downstream photogrammetric chain was unable to correct any radiometric defect of the optics. The digital image acquisitions using a CCD permit one to deal with these problems very correctly even with much cheaper and readily available materials, optimized for the quality of the picture, and not for typically photogrammetric aspects such as the distortion: these shortcomings are corrected by calculation. Remarks that follow sum up a few of these points:

- The field curvature is very inopportune, since it implies that the zone of better focusing is not plane: this defect must therefore be further reduced if one works with small pixels in the picture plane. For that reason it is necessary to pay attention to the parallel glass plates in front of the CCD sensor when one works with a large field, which is the rule in photogrammetry: this plate leads to an easily observable radial distortion, although not a troublesome one since it can be included in the calibration.
- The distortion has inevitably a symmetry of revolution. It is often described by an odd degree polynomial, and one should not be concerned by its importance, but rather by the residues observed between the polynomial and measures of calibration.
- The vignettage is governed by optics in the approximation of Gauss of a quite appreciable lighting loss of the sensitive surface following a law in $\cos^4$ of the angle between rays and the optic axis, which

reduces the brightness greatly within the angles of an image. This effect is very well corrected today, including optics for the general public, and can also be radiometrically calibrated in any camera, using an integrating sphere as a light source.
- The shutter is also an important piece. If it is installed close to the optic centre of the objective, its rise and fall times do not create a problem, as at each step of the shutter movement, all the focal plane receives the same light modulation, but then there is a shutter in each objective used, which increases their cost. For the digital cameras using some CCD matrixes, one will note that the very high sensitivity of these devices requires having access to a very short exposure time, which is not necessarily feasible in times of strong sunlight, considering the relative slowness of these mechanical devices. One is then sometimes obliged to add neutral density filters in front of the objective to limit the received flux, as it is impossible to effectively close the diaphragm, which would create excessive diffraction effects. Otherwise, if one uses a linear movement blade shutter with this type of camera, it is necessary to check that the scrolling of the window proceeds in the sense of the transfer of charges of the CCD. And the electronic compensation of forward motion (against linear blurring due to the speed of the airplane, often called TDI – see §1.5, that consists in shifting charges in the focal plane at the same physical speed as the scrolling of the picture in the focal plane) has to be activated during the entire length of movement of the blades (generally 1/125th of a second), and not during the time of local exposure of the CCD (that can range sometimes up to the 1/2,000th of a second). If the blade movement were oriented in another direction, it would indeed result in an important distortion of the image. And if its direction is good but the electronic compensation of forward movement is not activated at the correct time, there will only be a part of the image that will be correct, in sharpness and in exposure as well.
- Finally, on digital cameras, it is necessary to note that it is easy to calibrate the sensitivity of pixels individually, and to apply a systematic correction, which improves the linearity of the measure. Otherwise, in the matrix systems, the electronic compensation of forward motion leads to sharing of the total response over several pixels, which improves the homogeneity of the pixel response once more.

1.3 COLORIMETRY

Yannick Bergia, Michel Kasser

1.3.1 Introduction

The word 'colour' is a dangerous word because it has very different senses. We all have one intuitive understanding of this notion, and we think we know what is green, or orange. In fact, the word 'colour' regains two fundamentally different notions:

- on the one hand, the visual sensation that our eyes receive from the surface of an illuminated object;
- on the other hand, the spectral curve of a radiance fixing, for every wavelength, the quantity of light emitted or transmitted by an object.

The first meaning is eminently subjective and dependent on many parameters. The second is entirely objective: it corresponds to a physical measure. The fact that the language only proposes one term for these two notions is a source of confusion. We must always have to mind that our eyes only communicate to us a transcription of discerned light limiting itself to the supply of three values, and are not therefore capable of analysing spectral curves by a large number of points.

The objective of this presentation is to attract the reader's attention to the complexity of situations that will be met when some digital processes are to be applied, not to black and white pictures, but to pictures in colour. It will especially be the case, in digital photogrammetry, in processes of picture compression and automatic correlation. It is indeed certain that for such processes, it has been necessary since the conception of algorithms to take into account a given type of space colorimetry, and that it is necessary to choose it in functions, for example, of metrics problems that it presents.

Colours that our eyes are capable of differentiating, if they are not in infinite number, are nevertheless extremely numerous. To compare them or to measure them, one refers to the theoretical basic colours (often called primary) that are defined as follows. One divides the visible spectrum into three sub domains, for which one defines colours with a constant radiance level in all spectral bands:

- between 0.4 and 0.5 microns, the blue;
- between 0.5 and 0.6 microns, the green;
- between 0.6 and 0.7 microns, the red.

A light is said to be blue therefore (respectively, green, red) if it is composed of equal parts of radiances of wavelengths between 0.4 and 0.5 microns (respectively 0.5/0.6, 0.6/0.7 μm) and does not contain any other wavelength of the visible spectrum (one does not consider the remainder of the spectrum). One also defines, next to these 'fundamental' colours, basic colours called 'complementary':

- cyan, whose wavelengths range from 0.4 to 0.6 μm;
- magenta, from 0.4 to 0.5 μm and 0.6 to 0.7 μm;
- yellow, from 0.5 to 0.7 μm.

1.3.2 Trichromatic vision

The notion of trichromatic vision is one of the principles basic to colorimetry. It is based on the fact that the human eye possesses three types of receptors (cones) sensitive to colour (the eye also possesses other receptors, short sticks that are sensitive to brightness), each of the three types of cones having a different spectral sensitivity. This principle applies in such a way that a colour can be represented quantitatively by only three values. Besides, experiences prove that three quantities at a time are necessary and sufficient to specify a colour. Also, they introduce notions of additivity and proportionality of colours and so confer the intrinsic properties of a vectorial space of dimension three to all spaces where one would try to represent colours.

1.3.3 Modelling of colours

To specify a colour means to associate with the spectrum of distribution of energy of a given light a triplet of values that will represent it in what will be then a colorimetric space. Considering the results of experiences, one can consider that providing three colours respecting conditions of mutual independence (and that will be then the three primary axes of the space) can be sufficient to define a colorimetric space: for a given colour, multipliers coefficients of the three intensities of sources associated to primaries (when the reconstituted colour corresponds to the considered colour) will be coordinates of the colour in the constructed space. Another way to express this is to return to the physical data that contains the colour information and to integrate the spectral distribution of optical power on the visible wavelengths while using three functions linearly independent of weighting to define three channels. Thus, one can construct a colorimetric space, for example the fictive ABC system, by defining three functions of weighting $a(\lambda)$, $b(\lambda)$, $c(\lambda)$ (see Figure 1.3.1). Then every transformations, linear or not, of such a colorimetric space will also be a colorimetric space. There are an infinite number of such spaces, but one will find – in what follows – only certain spaces are often met with.

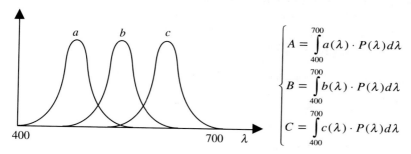

Figure 1.3.1 Definition of a fictive colorimetric reference.

1.3.4 Survey of classic colorimetric spaces, the RGB, the HSV and the L*a*b*

1.3.4.1 The RGB space

This is the most often met colorimetric space, used in particular for television and computer screens, of so-called 'additive synthesis' (as opposed to the 'subtractive synthesis' while using white light and subtracting from it a part of the spectra by filters, which is for example the case in printing, where each ink acts like a filter on the white paper). Its definition rests on the choice of three primary colours: the red, the green and the blue.

Considering data processing, the RGB space is represented as a discretized cube of 255 units of side. All feasible colours for computer applications are represented therefore by a triplet (R, G, B) of the discrete space. We may note the first diagonal of the reference frame where colour is distributed with the same value for the three channels, R, G and B: it is the axis of the grey or achromatic axis.

This space does not present specific points, but it is the one that the current material environment imposes on us: the large majority of pictures in true colours susceptible of being processed is coded in this system. We will also present methods of conversion from RGB towards another very common system, the HSV space, this only as an example.

1.3.4.2 The HSV space

Definitions

The Hue, the Saturation and intensity Value that constitute the three components of this system are the relatively intuitive notions that allow a better description of the experience of the colour. Some more theoretical definitions can be given.

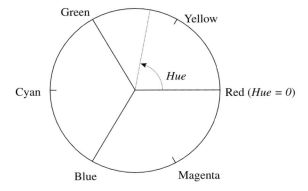

Figure 1.3.2 Distribution of basis colours on the circle of hues.

The intensity can be presented as a linear measure of the light power emitted by the object that produces the colour. It allows the dark colours to be distinguished naturally from the bright ones.

The hue is defined by the CIE (Commission Internationale de l'Éclairage) as the attribute of the visual perception according to which a zone appears to be similar to a colour among the red, the yellow, the green and the blue or to a combination of two of them. It is a way of comparing known colours on which it is easy to agree. According to this data, colours can be represented on a circle. The red, the green and the blue are there with an equal repartition (the red having by convention a null hue). Figure 1.3.2 illustrates this distribution.

The saturation allows the purity of a colour to be determined. It is while making the saturation vary from a null hue colour, for example, that one will move from the pure red (most saturated), that can be placed on the circumference of the hue circle, toward the neutral grey in the centre while passing all pink tones along the radius of the disk. From a more physical point of view one can say that if the dominant wavelength of a spectrum of power distribution defines the hue of the colour, then, the more this spectrum will be centred around this wavelength, the more the colour will be saturated.

Considering these definitions, colours specified in the HSV space can be represented in a cylindrical space and are distributed in a double cone (see Figure 1.3.3).

Distance in the HSV space

To be able to exploit this space for process applications it is necessary to define a distance. Naturally, the Euclidian distance does not have any sense here, the obvious non-linearity of its cylindrical topology and in particular the discontinuities met in the neighbourhood of the axis of inten-

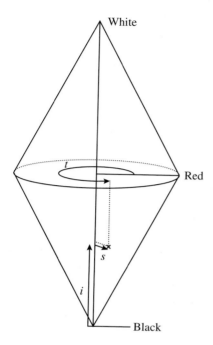

Figure 1.3.3 Theoretical representation of the HSV space.

sity and the passage of the null hue are unacceptable. One has thus agreed to consider as a distance between two points in this duplicate cone the length of the smallest bow of a regular helix that it is possible to draw between them.

Let us consider two points of the HSV space therefore, P_1 (I_1, T_1, S_1) and P_2 (I_2, T_2, S_2). One stands in an orthogonal Euclidian reference frame whose vertical axis (Z) is confounded with the axis of intensities, and the first axis (X) is directed by the vector of hue of the point whose intensity is smallest (suppose that it is P_1). One notes ΔI the positive difference of intensity between the two points (in our case $\Delta I = I_2 - I_1$), ΔS the difference of saturation ($\Delta S = S_2 - S_1$) and θ the smallest angular sector defined by the two angles of hues.

We are then in the situation illustrated by Figure 1.3.4 showing the bow of helix for which we need to calculate the length precisely. In these conditions we propose the following parametric representation of this bow:

$$P(t) = \begin{cases} X(t) = S_1 + t\,\Delta S)\cos(t\theta) \\ Y(t) = (S_1 + t\,\Delta S)\sin(t\theta) \\ Z(t) = t\,\Delta I \end{cases}$$

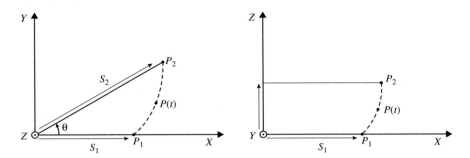

Figure 1.3.4 Portion of helix to compute distances in the HSV colorimetric system.

The length of the bow is then given by the calculation of the curvilinear integral

$$D_h = \int_0^1 \left\| \frac{\partial P(t)}{\partial t} \right\| dt$$

that provides the following result:

$$D_h = \frac{1}{2\,\Delta S}\left[\left(S_2\sqrt{\Delta S^2 + \Delta I^2 + S_2\theta^2} - S_1\sqrt{\Delta S^2 + \Delta I^2 + S_1\theta^2}\right) + \frac{\Delta S^2 + \Delta I^2}{\theta}\left(\operatorname{argsh}\left(\frac{S_2\theta}{\sqrt{\Delta S^2 + \Delta I^2}}\right) - \operatorname{argsh}\left(\frac{S_1\theta}{\sqrt{\Delta S^2 + \Delta I^2}}\right)\right)\right].$$

(1.31)

Technical way of use of the HSV space

There does not exist a unique HSV space strictly speaking, rather numerous definitions or derivative specifications. Indeed several quantities deduced for example from the R, G and B values of a specified colour in the RGB space can correspond to very subjective definitions of hue, intensity and saturation that have been given. Conceptually all these spaces are very much the same in the sense that they all allow one to easily specify a sensation of colour in terms of hue, intensity and saturation, or if need be neighbours and equivalent notions.

The most often used definition is that found for example in Gonzales and Woods (1993): the calculation takes place on the r, v and b values between 0 and 1 (it is about the R, V and B initial values divided by 255). If $r = v = b$, the considered colour is a nuance of grey, and one imposes by convention: $t = s = 0$; it does not affect the reversible character of the transformation.

$$i = \frac{r + v + b}{3},$$

$$t = \begin{cases} \arccos\left(\dfrac{(r-v)+(r-b)}{2\sqrt{r^2+v^2+b^2-rv-vb-br}}\right), & \text{if } v \geq b \\ 2\pi - \arccos\left(\dfrac{(r-v)+(r-b)}{2\sqrt{r^2+v^2+b^2-rv-vb-br}}\right), & \text{if } b \geq v \end{cases} \quad (1.32)$$

$$s = 1 - \frac{\min\{r, v, b\}}{i}.$$

1.3.4.3 The L*a*b* space

L*a*b*, defined by the CIE in 1976, has the goal of providing a space of work in which the Euclidian distance has a true sense: it develops a space uniform to the perception, that is to say in which two greens and two blues separated by the same distance would lead to the same sensation of difference for the human eye, which will not be for example the case in

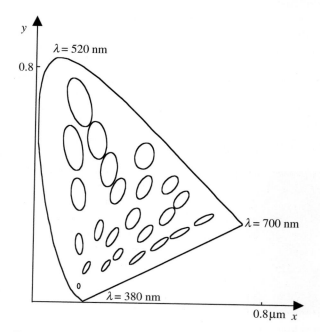

Figure 1.3.5 Separation of colours in the chromatic L*a*b* diagram.

the RGB system basis as we have a better capability to differentiate in the blues than in the greens. This affirmation has been verified by the experience of the chromatic matching of Wyszecki (Wyszecki and Stiles, 1982) whose results are presented on the chromatic diagram of Figure 1.3.5. Ellipses that are drawn there (whose size has been exaggerated nevertheless) represent, according to the position in the diagram, zones whose colours cannot be discerned by observers.

The transformation toward L*a*b* will distort the space so that these ellipses become circles of constant radius on the whole of the diagram. It is this important property of linearity of perception that forms its essential interest: if one works in a metrics that replicates mechanisms of the human vision, one can expect that the result of segmentations obtained be relatively satisfactory, in the sense that it will correspond more to the delimitation that a human operator could have drawn by hand while seeing the picture.

1.3.5 Distribution of colours of a picture in the different spaces

In order to place this survey of colorimetric spaces in the context of the picture process, we have found it interesting to represent a picture in these different spaces. Naturally, whatever is the space in which the picture is coded, its aspect must remain the same, but not the distribution of its colours in the three-dimensional space of colours representation. One considers the aerial photograph of an urban zone as an example (see Figure 1.3.6). And one gives the representation of the colour clouds that are present in several classic colorimetric spaces, including those seen previously (Figure 1.3.7).

This horizontal comparison imposes an observation: if the cloud of points keeps, with the exception of the orientation, the same aspect in the two spaces linearly deriving from RGB (KL and XYZ), its geometry in the two uniforms spaces of the CIE (L*a*b* and L*u*v*) is radically different. The change of space has consequently a deep modification of the metrics

Figure 1.3.6 Example of colour image used to test various colorimetric spaces, here displayed in black and white.

Colorimetry 33

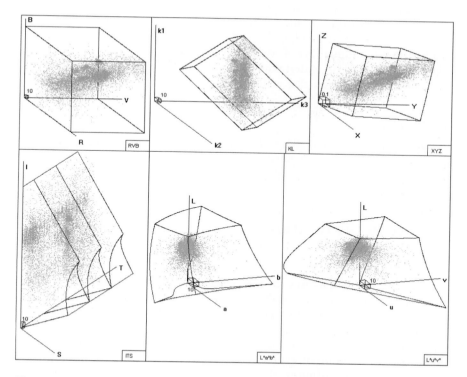

Figure 1.3.7 Distribution of colours of the picture in the different colorimetric spaces.

that allow one to anticipate very variable results for the automatic process applications according to the chosen workspace.

1.3.6 Conclusion

It is therefore important not to underestimate problems due to the use of colour in digital imagery, in particular for photogrammetric processes. At least one should know how to exploit the passage of a space to one dimension (the black and white) to a space of at least three dimensions. It is certain that in a few years there will be for the disposition of users some spatial images with small pixels of interest to photogrammetrists, and using more than three spectral bands. Processes used for the black and white will be reused with multi-spectral pictures, but it will be at the cost of a considerable loss of information. In fact it will be necessary to start from zero systematically the analyses of the studied phenomenon to extricate the best party possible of the available spectral bands. In this optics, one will need to adopt a colour space with a metrics adapted to the problem each time: compression of the image, automatic correlation, automatic segmentation, automatic shape recognition, etc. Here we presented some details of trichromatic colorimetry, because it was necessarily very well studied for the human eye. It is necessary to note therefore that an equivalent study should be undertaken for a different number of spectral bands, at each time where the case will happen. . . .

References

Wyszecki G., Stiles W.S. (1982) *Colour Science: Concepts Methods and Quantitative Dated Formulae*, second edition, John Wiley & Sons.

Gonzales R.C., Woods R.E. (1993) *Digital Picture Processing*, Addison-Wesley.

1.4 GEOMETRY OF AERIAL AND SPATIAL PICTURES

Michel Kasser

The usable digital pictures in photogrammetry may originate from three possible sources: digitized traditional silver halide pictures, digital pictures originating from matrixes or airborne linear CCD cameras, and the spatial images of high and very high resolution (between 10 m and 0.8 m pixel size), these also constituted of CCD matrixes or linear CCD. The geometry of these different types of pictures is very different, and will be presented here according to their specificity.

1.4.1 Images with a large field and images with a narrow field

What distinguishes pictures of spatial origin from those using an airborne sensor is only the angular field of view, obviously narrower in the case of the first ones. This field has an influence on several aspects of images, in particular:

- The direction of observation is pretty much constant in relation to the solar radiance for the spatial sensors, whereas it varies considerably in the case of pictures with a large field. Thus we have the evidence of a phenomenon of brilliant zone, called 'hot spot', which is due to the fact that when the direction of observation comes close to the direction of the solar radiance, one can no longer observe any zone of shade whatever the roughness of the ground, or the vegetation, or the buildings. This direction has a very weak probability to be met in an image with a narrow field. On the other hand one very often meets it in aerial pictures with a large field as soon as the sun is high on the horizon. It materializes a zone of the picture where the reflectance of the ground becomes much higher than elsewhere. Another important point: the direction of reflection of the sun on surfaces of free water is, for the same reasons of size as for the field of vision of the sensor, susceptible of provoking a prompt blinding of sensors having a large field, a situation that is also statistically not frequent on the spatial high-resolution imaging devices.

- The phenomena of refraction: the light rays captured by a spatial imager are practically all parallel (the angular field of the Ikonos satellite, launched in 1999, is 0.9°), and phenomena of differential refraction between the extreme rays are very weak, otherwise negligible. This refraction has only very few effects on the absolute pointing of the sensor in the space, since it leads to a curvature that is significant only in the last kilometres close to the ground, with therefore a negligible angular effect since the satellite is between 500 and 800 km away at least. These aspects are quite different for the airborne imagers with a large field of view (see §1.1).
- On the other hand, for the phenomena bound to the atmospheric diffusion (see §1.2), there is no difference between an image obtained from an airborne sensor flying to an altitude higher than 4 or 5 km and an image obtained on board a satellite, since in the two cases all the diffusing layers have been crossed.
- The stereoscopic acquisition on a plane is performed in a very simple way, essentially because the airplane can acquire any images only when it flies on a straight axis. According to the requested longitudinal overlap rate and to the speed of the plane in relation to the ground, one chooses a rate of image acquisition. And the axis of any image is in all cases practically vertical. On the other hand, on a satellite one can get a stereoscopy in several modes, along the track or between two different orbits. When one works in stereoscopy along the track (so-called 'ahead–rear' stereoscopy, between two similar instruments pointed ahead and behind, or with an instrument capable of aiming very quickly in the direction of the orbit), considering the speed of the satellite on its orbit (of the order of 7 km s^{-1}), pictures are nearly exactly synchronous. In this case the direction of the stereoscopic base is by necessity that of the track of the satellite, that is imposed by laws of celestial mechanics and that is virtually impossible to change. But from several different orbits the satellite can also aim at the same scene, which provides a 'lateral' stereoscopy, and then a considerable time difference between acquisitions may happen, which sometimes complicates the work of photogrammetric restitution: for example the position of the sun is different and therefore shades changed a lot (this induces difficulties with the automatic correlators), vegetation changed (new leaves, or loss of leaves as well), cultures are completely different, levels of water in streams, lakes, or even the sea (tides) are no longer the same, vehicles have moved, etc. . . . to acquire pictures as quickly

as possible, satellites are obliged to use the lateral pointing as well as the ahead–rear stereoscopy, but it is necessary to understand all the consequences of this for the acquired images when one considers photogrammetric processes.

1.4.2 Geometry of images in various sensors

There are essentially two main geometries of sensors that are used for photogrammetric processes: the conical geometry and the cylindro-conic geometry.

1.4.2.1 Conical geometry: use of films or CCD matrixes

This is well known, since it is the main means of human vision that has been used since the invention of photography by Nièpce or of the cinema by the Lumière brothers. It essentially implies a sequential mechanism to acquire a picture: a very brief exposure time, followed by one technically necessary dead time to restock a virgin length of film, or to transfer and to store the digital values corresponding to every pixel on a digital camera. This dead time is profitably taken on satellites to reorient the aiming system in order to image as many possible scenes in relation to the plan of work that is requested.

If the necessary exposure time is too long, there occurs a phenomenon of linear blurring, because the picture moves continually in the focal plane with respect to the movement of the airplane or the satellite, which transforms the picture of a point into a small segment. To overcome this, the correction of linear blurring (forward motion compensation, or FMC) on cameras using silver halide films is performed by moving the film during the exposure time. This film is pushed, indeed, by depression on a very plane surface (the bottom plate of the camera), and a mechanism moves this plate in order to follow exactly the movement of the image in the focal plane during the short period of opening of the shutter. For the CCD matrixes it is even simpler; it is sufficient to provoke a charge transfer at the proper speed so that a given set of charges is created by only one elementary point of the ground: as the picture of this point moves, one will make the charges created move at the same speed. The FMC, in these two cases, requires absolute knowledge of the speed vector of the image in the focal plane, as well as a general mechanical orientation of the sensor so that its compensation capacity can be parallel to this speed vector. It is not necessarily easy to know this speed vector in an airplane, since the leeway of the plane under the lateral wind effect must be known also. Evidently these parameters of speed may be advantageously provided by the system of navigation of the plane, currently the GPS. And on a satellite, these parameters are deduced directly from the orbit and from coordinates of the target.

1.4.2.2 Cylindro-conic geometry: use of a linear CCD

This geometry is well known in remote sensing (airborne radiometers, Landsat satellites ...), and has been used also in photogrammetry because of the Spot satellites without discontinuity since 1986. This geometry is provided naturally by every linear sensor (linear CCD) placed in the focal plane perpendicularly to the speed vector of the image. It is therefore the movement of the plane or the satellite that allows one of the dimensions of the image to be described. The geometric details of these sensors used in photogrammetry rely especially on the more or less stable speed vector.

In a satellite, the trajectory is locally very precisely a Keplerian orbit, there is no lack of stability to fear on the parameters of position. On the other hand the satellite is frequently equipped with solar panels that are often reoriented automatically thanks to small motors, which create, in backlash, small changes of attitude. And generally it is impossible to guarantee a perfect aiming stability during the acquisition of an image, which results in a small distortion, that may be corrected in the most advanced means to process the geometry of the images.

But on a plane, the small movements in roll, pitch and yaw are permanent, with an amplitude and a frequency that depend directly on the turbulence of air. Therefore inevitably one must perform a permanent record at least of the attitude, on the three axes, of the sensor. In fact one has also to measure permanently the three coordinates, as precisely as possible, of the optic centre of the sensor, using an inertial system and a precise GPS receiver whose data are quite complementary. Once the data are acquired one may reconstitute the images as one would have acquired them if the trajectory of the plane had been perfectly regular. But this correction, in spite of the efforts performed using excellent inertial platforms, does not succeed to perfectly correct these movements as soon as there is a little turbulence, which is very often the case for the urban surveys, generally done at a low altitude and therefore in a turbulent flight zone. This results in edges of buildings that are not straight, and finally residual image distortions that can often cover several pixels. In particular, there may be real hiatuses of data, owing to the angular movements of the plane being too rapid, and on another hand geometric aliasing of the picture, a unique point being imaged on several lines. Such discontinuities are sometimes impossible to correct by computation, and the only possible resource is then to 'fill' these hiatuses by interpolations between neighbouring strips, but this is not effectively very satisfactory in terms of picture quality. The only possibility to avoid such situations consists in installing the sensor on a servoed platform correcting in real-time movements of the plane, which is obviously a considerable overcost for the equipment.

In counterpart, this technology permits data coming from several linear CCD to be combined, with potentially no limit to the number of pixels measured along a line.

In this cylindro-conical geometry it is necessary to note a point that is absolutely fundamental in photogrammetric uses of the acquired images. Contrary to what occurs in a conical geometry, points are never seen on several consecutive pictures, since in fact every line of data is geometrically independent of the previous one. Thus the exact knowledge of the trajectory is mandatory to be able to put in correspondence the different successive groups of points seen by the linear CCD. The trajectography is therefore an essential element of geometry allowing the reconstitution of the relief, and the images alone are not sufficient to proceed to a photogrammetric restitution. The only possible validation of these parameters of trajectory, which become so important here, consists in using effectively three directions of observation in order to pull a certain redundancy of determination of the DTM (digital terrain model), that may validate the whole set of parameters introduced in the calculation. This point is fundamentally different from the usual geometry with the traditional cameras, since the stereoscopic basis doesn't need to be known then (even if we find an appreciable economical interest in the use of a GPS receiver in the plane, and even an inertial platform too, in order to get the absolute orientation of the images, and thus to avoid the determination of ground reference points, always quite expensive to get; but this equipment is not mandatory), and that it is in any case redetermined by the calculation when the model is formed.

Pictures obtained by such devices can be acquired in very long strips, continuously if need be, which suppresses worries of splices between images in the longitudinal sense, parallel to the axis of flight. The carving in successive images is therefore artificial and can be modified by the user without any problem. Here we should note also that a very strict calibration of the radiometric response of different pixels is mandatory, as it is for the linear CCD scanners used to digitize aerial pictures; any failure to complete it will result in unpleasant linear artefacts. For that reason the Ikonos satellite has included on board a calibration device regularly activated, that exposes all pixels to the solar light.

As for the matrix systems, the stereoscopy can be used in the 'ahead–rear' mode but also, as for satellites, by using lateral aiming (case of the Spot satellites). The airborne systems follow the principle of stereoscopy 'ahead–rear' and are described in §1.5.2.

The cylindro-conical geometry requires algorithms that are quite different from those used in classic conical geometry for photogrammetric restitution. These algorithms have been explored extensively for the restitution of the Spot images since the 1980s. Nevertheless, they remain poorly known and forbid de facto the use of equipment and software not designed for this geometry, which thus prevents the use of these data on other stereo-plotting devices.

1.4.3 Conclusion

The different geometries used to acquire pictures have some important consequences when a photogrammetric process is to be performed. It is often not just a detail, since if one doesn't take the necessary precautions one may find it impossible to validate by internal means the results obtained. Much attention must be paid to the practical consequences of choices of images, in particular with the arrival on the market of very high spatial resolution images, apparently very similar to the aerial pictures, but which have completely different geometric properties.

In the domain of the large field airborne imagery, linear CCDs systems seem especially destined to provide very high resolution pictures, but whose metrics is not very critical (devoted to photo-interpretation, for example). On the other hand the matrix systems, providing an extremely rigorous geometry, will be much more suitable to photogrammetric uses.

1.5 DIGITAL IMAGE ACQUISITION WITH AIRBORNE CCD CAMERAS

Michel Kasser

1.5.1 Introduction

Currently most processes of photogrammetry turn toward a larger and larger use of digital images. Considering the reliability and the quality of materials for aerial image acquisition used several decades ago, but also considering the excellent knowledge that we have about the defects of this chain of image acquisition, research laboratories and constructors wisely started with the reuse of previous achievements. They proposed therefore first to digitize images obtained in aerial image acquisitions on argentic traditional film. Nevertheless, this solution is only a temporary solution, and as we will see in §1.8, the digitization of films is a delicate and costly operation, which brings its own share of deterioration to images. It is therefore natural that studies have been led, initially, since 1980, at the DLR in Germany for the Martian exploration, then in France since 1990, to get digital images directly in the airplane itself. Studies moved in two simultaneous directions, using two CCD devices: linear CCD, according to the model of sensor popularized by the spatial imagers like SPOT or Ikonos, using the advancement of the plane to scan one direction of the image, and matrixes, that are the modern and conceptually simple copies of traditional photography.

These two directions of research first gave rise to realizations of laboratories (e.g. HRSC sensor using several linear CCDs of the DLR, matrix cameras of the IGN-F), then to industrial realizations (ADS 40 of LH Systems in 2000 and MDC of Zeiss Intergraph in 2001). Studies have

Figure 1.5.1 (a) The digital modular camera DMC 2001 from Zeiss-Intergraph.

Figure 1.5.1 (b) Three digital cameras from IGN-F: (from left to right) with one (1995), with two and with three separate cameras (2000), using KAF 6300 (2k × 3k pixels) or KAF 16800 (4k × 4k), or KAF 16801 (4k × 4k with antiblooming) chips from Kodak.

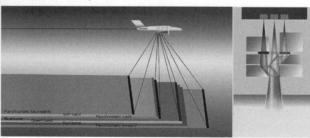

Figure 1.5.1 (c) The ADS-40 digital camera from Leica-Helava (left), the CCD lines observing various wavelengths (centre) and the optical subsystem for the separation of the visible light into 3 colours RGB (right).

Digital image acquisition 41

shown quite clearly fields of interest of these two techniques, as well as their limitations. They have also shown how much the community of photogrammetrists had difficulties imagining the future with tools very different from older ones: this community is ready to replace one link of the photogrammetric chain at a time, but not much more, which is easily comprehensible. Photogrammetrists have, for example, clearly shown their concern, not because of directly digital images, but because of the entirely new situations where their empirical ability would have to be reconstituted from zero, in particular being entirely at the mercy of software editors. For example, linear CCD systems give a quite different geometry to process the images as compared to traditional techniques.

The digital cameras developed according to these two general types are presented briefly below. They have in common the use of the CCD, which offers an extraordinary luminous response linearity compared with the usual argentic processes: these are really radiometers, perfectly calibratable, that make a much easier job, for example at the radiometric connections between neighbouring images at the time of the realization of mosaics or orthophotographies. The advantage of this linearity is also very clear when one has to suppress by digital processes the atmospheric diffusing fog: this operation is extremely simple and efficient with the CCD sensors, whereas it is impossible to automate it on argentic colour images.

Currently the two solutions (linear CCD and matrixes) seem to have slightly different domains of applications, and as we already evoked it in §1.4, linear CCD systems are preferably destined to provide images of very high resolution but whose metrics is not very critical (for the merely visual exploitation and without precise measurements, for example, which represents a very large part of the market of aerial imagery). On the other hand the matrix systems, providing an extremely rigorous geometry (and even much better than that of argentic pictures, digitized or not), are far more suited to photogrammetric uses. But only after several years of market reaction to these offers can we make valid conclusions, and it is not possible for the authors to push the analysis farther at the time of writing this book. Some elements already presented in §1.4 will not be repeated here, and we refer the reader back to this section if necessary.

1.5.2 CCD sensors in line

It was in order to prepare a mission to Mars, that finally didn't fly, that the DLR (Germany) studied in detail the HRSC tool (high resolution stereo camera). This tool, which was conceived for the spatial applications, was then tried with success on airborne missions. It has been used industrially since 1999 in production within the Istar company (France), and the new camera ADS 40 of LH Systems is based on the same concepts.

The principle is as follows: in the focal plane of high-quality optics is installed a set of linear CCDs, which allows at any given instant the ground to be imaged in several directions. The prototype of the ADS 40 (Fricker, *et al.* 1999) uses four linear CCD or groups of linear CCD of 12,000 elements each, one that gets images to the vertical of the plane, with the others imaging following a fixed angle with respect to the vertical (several angles are possible, according to the uses) toward the front and the rear of the plane. Thus, a point of the ground is seen successively under three different angles, which quite obviously permits one stereoscopic reconstitution of the relief. Some of these sensors work in panchromatic mode, others are equipped with spectral filters. Some linear CCD used in panchromatic mode are formed of two staggered linear CCD, separated by half a pixel, which may allow one to rebuild the equivalent of the signal provided by a linear CCD of 24,000 pixels. One then has a resolution that is without any comparison with any other airborne sensor used, even though the geometric quality of such images allows some minor distortions.

Considering the trajectory of the plane, that is evidently uncertain and subject to unforeseeable and important movements, it is necessary to know at any instant and with a very high precision the real position of sensors in the space. This problem, which does not appear on a satellite (whose trajectory is extremely steady), implies for a plane the use of powerful additional systems capable of correcting at different times the effects of movements of the plane on the recorded image, that would be completely unusable without it. For that purpose, one attaches to the system an inertial platform as well as a GPS sensor; the fusion of the output data of these two sources is made exploiting a Kalman filter. Let us note that the coupling inertial platform + GPS doesn't necessarily permit one to determine in a satisfactory way the yaw movements of the plane: it would be determined correctly by the Kalman filter only if the plane did not remain for a long time right in line, which is unfortunately unavoidable. And long-distance systematisms of the GPS (for example, if it is used in trajectography mode), are directly transferred as deformations of the survey performed. Besides, one should not overestimate the operational performances of the GPS, and very short losses of signal can occur in flight, which leads at the time of the treatment to short interruptions of the corrections provided by the inertial platform: it almost produces automatically a set of irretrievable distortions of the image acquired. Therefore it is not foreseeable to get a geometric precision, expressed in pixels, comparable with what the digital images usually provide with matrix imagers in conical perspective. Typically one can count, in equivalent cases, on residual errors of several pixels for images obtained with linear CCD sensors, against 0.1 pixels on matrix images.

The results published in 2000 show that the pros and cons of this sensor type ADS 40 include typically:

- The possibility of an image with a number of distinct points on the ground close or even superior to what is available in aerial digitized pictures (12,000 pixels are roughly equivalent of what provides an

aerial picture 24 × 24 cm digitized with 20 μm of pixel, and 24,000 is achievable if necessary).
- The geometric distortions due to inertial system insufficiencies are weak, but remain generally visible (some pixels): the metrics of the image therefore shows some small defects.
- The dynamics of the image is satisfactory, but the necessity to transfer charges in the linear CCD to a very high speed implies a very limited exposure time, which does not allow dynamics as high as that on matrixes to be reached. Images, nevertheless, offer an excellent linearity of response.
- The geometry of images (cylindro-conic perspective) implies new modes of digital data processes, which seem quite suited to the automated determination of digital surfaces models (DSM), but forbid the use of standard softwares of classic photogrammetry stations.
- The restitution of the colour is obtained by adding in the focal plane additional linear CCDs equipped with filters. For example, the linear CCD that images along the vertical is in fact replaced by three parallel and very close linear CCDs equipped with the three usual filters, RGB. Other linear CCDs can be equipped with other filters, e.g. for the infrared. But if the three colours are obtained through lines that do not observe exactly the same area simultaneously, some artefacts appear on mobile objects, of course, but also when strong perspective differences are present in the image (high buildings, for example).

1.5.3 Matrix CCD sensors

The principle consists in preserving the usual geometry of the classic image acquisition (conical perspective), and using a CCD matrix as large as possible in the focal plane of the optics. Studies performed at IGN-F from the beginning of the 1990s (Thom and Souchon, 1999) show how such images could be inserted in photogrammetric classic chains, and what were the limitations and advantages of such matrix cameras:

- The CCD matrixes are perfectly suitable for the compensation of linear blurring due to the speed (forward motion compensation or FMC), and even with more simplicity than on traditional FMC airborne cameras since it is performed electronically, without any mechanical movement. This allows a long exposure duration (several tens of milliseconds if need be), compatible with a very weak lighting (wintry or dusky image acquisitions) while preserving an excellent signal/noise ratio.
- The sensitivity of the CCD, added to the possibility of using long exposure times, is the origin of a dynamics of the image that can reach considerable values, a digitization on 12 bits being hardly sufficient. It is quite different from silver halide film image performances that, even if digitized in the best conditions, would hardly deserve 6 bits.

It is an essential asset, for example, for removing easily the atmospheric fog in post-process, or for observing in a satisfactory way shades and in the very luminous zones as well. It is also an essential asset for tools of automatic image matching, since with such a dynamic one no longer finds any uniform surface, and numerous details with quite faint differences of radiometry remain discernible.

- The largest commercially available matrixes in 2000 have dimensions of the order of $7,000 \times 9,000$ pixels. This represents a more modest pixel number than that of a digitized silver halide image. Nevertheless, cases of the use of such images of very high dynamics have been valued on applications of current middle-scale cartography: an equivalent service to a given ground size of pixel on a digitized argentic image is more or less provided by a two times larger pixel on a digital image of very large dynamics. Compared to the service provided by a traditional digitized image, it corresponds, therefore, to a digital image about half as large. Let us note that with the remarkable performances of these images it would be quite acceptable and logical to interpolate pixels of dimension half as large, which would so give the equivalent of an image $8,000 \times 8,000$ from a matrix $4,000 \times 4,000$. One would thus get the content of an image quite similar to a digitized argentic one, although of incomparably better linear radiometry. Nevertheless, for a given site of study this may imply, according to the chosen matrix size, more axes of flight than for a classic image acquisition, which is a source of additional costs. Developments of large dimension matrixes, by chance, follow the considerable demand for digital pictures for the general public, so that they evolve rapidly. Apparently formats used by amateurs (24×36 mm) and professionals (60×60 mm) are currently well covered by matrixes of $2,000 \times 3,000$ and $4,000 \times 4,000$. The likely evolutions could go circa a strong reduction of costs, the trichromy, and maybe an increase of current matrix sizes, but not necessarily very important (rather for the professional photographers, who represent a small section of the market) knowing that the data storage is no longer a limitation for these markets due to the availability of large data storage capabilities. The way chosen by the ZI society (DMC camera) consists in a combination of various modular matrix subsets based on matrixes about $4,000 \times 7,000$, capable of reconstituting $8,000 \times 14,000$ images, with the possibility of supplementary matrixes equipped with filters capable of reconstituting of colours. The elementary matrixes are equipped with individual optics aiming to divergent axes and assuring a small overlap between each of the four images. The resulting geometry is equivalent, within an accuracy of about 0.1 pixels, to an image acquired in only one block, by one optic.
- The restitution of the colour can be obtained of two ways: either on certain matrixes a set of filters is deposited on the pixels, or one uses

three matrixes equipped separately with filters. The first solution induces delicate problems of reconstitution of the colour, and requires well-developed software to let only small artefacts (e.g. on borders, or on very local details displayed on only one pixel). It implies either a subset of 4 pixels (e.g. Kodak: two with green, one with blue and one with red filters), or a different arrangement of the matrix, where the lines are staggered from a half-pixel, and the 3 filters are regularly spaced on a 3-pixel subset (e.g. Fuji: on each line the filters are R G B R G B ..., for example the R being staggered by 1.5 pixels from a line to the other), a solution that provides a better colour synthesis. The second solution is costlier, because it finally results in the equivalent of three complete cameras regrouped together, but it permits the perfect reconstitution of colours in all circumstances.
- The metrics of the image is absolutely excellent, and even significantly superior to the accessible metrics on classic film cameras. Even if the latter are nearly perfect in terms of distortion, the film represents a medium whose distortions are far from negligible. These distortions are due to treatments at the time of the development and the drying, that stretch the film in an anisotropic way. They are also, and this phenomenon is not so well known (see §1.1), due to dust on the bottom plate of the camera that prevents a good contact of the film on the table to depression, and creates some significant bumps at the time of the image acquisition. It provokes local distortions of the image that are pure artefacts, nearly impossible to model. In the digital matrix cameras, the calibration is performed as for a classic camera, but as the sensor is monolithic, no distortion can now occur, and one observes a precision in geometric computations that is generally better that 0.1 pixels.
- The available optics offers a much wider choice for amateur cameras than for the classic aerial cameras. Practically all optics adapted to the 60×60 format (professional photography) are satisfactory, and this at obviously much more modest costs. Indeed the classic defects of optics, distortion, vignetting, etc., can be corrected by a posteriori calculation, and only the resolution of the optics and its dimensional stability in the time remain important parameters.

1.5.4 Specifications of digital aerial image acquisitions

The use of airborne digital cameras induces significant changes in terms of specification of image acquisitions. Notions of height of flight and size of pixel can be considered as largely independent, in particular due to the much larger choice of available optics. Even the notion of scale disappears completely, with all the empirical knowledge that is attached to it. *The maximal angle of field* of the image will be chosen according to the acceptable hidden parts and the need of precision on an altimetry restitution, expressed in particular through the B/H ratio of the distance between successive positions of the camera at the time of two successive image acquisitions, divided by the altitude of flight. *The size of pixel* will be chosen according to the size and the nature of objects to survey and to detect, and to the dynamics of the accessible image with the considered camera. It is with these two parameters that one will determine the project of flight, and in particular the altitude.

One will note some empirical rules that will be detailed in §1.9: the precision of pointing of an object is performed in general to better than 0.3 pixels, and to identify an object easily it is desirable that it is not smaller than 3×3 pixels. This allows one to specify the really necessary pixel sizes. For example, with a matrix camera with a very large dynamics, to achieve an orthophotography on an urban zone where a DTM is already available, one will choose a long focal distance to limit the hidden parts of buildings, and the raw size of pixel will be chosen, e.g. as 50 cm to have a satisfactory description of buildings, or of 30 cm if one wants to have a tool of management of urban furniture. These sizes of pixels will be reduced if the dynamics of the image is low (respectively 30 cm and 15 cm for a digitized classical image).

In a traditional image acquisition the height of flight also intervenes by the atmospheric fog, more or less acceptable, that will result: sometimes, to get a very weak fog with a height superior to 4,000 m may prove to be an impossible mission under certain latitudes in classic photography. In digital imagery, if the dynamics is sufficient it allows the removal of the atmospheric fog while preserving a good quality of colour and a good capacity to work in zones of shades. It permits a considerable reduction in the delays bound to the meteorological risks, but also eases the constraint on the parameter of flight height, if need be.

References

Fricker P., Sandau R., Walker S. (1999) Digital Aerial Sensors: possibilities and problems, *Proceedings of OEEPE Workshop on Automation in Digital Photogrammetric Production*, Marne-la-Vallée, 22–24 June, Public. OEEPE no. 37, pp. 81–90.

Thom C., Souchon J.-P. (1999) The IGN Digital Camera System, *Proceedings of OEEPE Workshop on Automation in Digital Photogrammetric Production*, Marne-La-Vallée, 22–24 June, Public. OEEPE no. 37, pp. 91–96.

1.6 RADAR IMAGES IN PHOTOGRAMMETRY
Laurent Polidori

1.6.1 Overview

Images obtained from a synthetic aperture radar (SAR) are very sensitive to terrain undulations, and this sensitivity has been the basis for the development of several relief-mapping techniques using one or several radar images, namely, radargrammetry, interferometry and radarclinometry (Polidori, 1991). Each of these techniques was initially proposed more than 25 years ago, and their theoretical bases were established in the 1980s. The state-of-the-art of radar mapping has been presented in detail by Leberl (Leberl, 1990), based on airborne campaigns and on pioneering spaceborne SARs like SEASAT (1978), SIR-A (1981) and SIR-B (1984), but the actual potential of radar techniques has been known only for a few years, due to the lack of SAR data before the 1990s. The launch of several spaceborne SARs in recent years (ERS, JER-S, RADARSAT, SIR-C) has led to further experiments and to a better estimation of the accuracy of radar-derived DSMs (digital surface models).

Several specific properties of radar sensors must be mentioned to explain their potential for mapping.

An active device

A radar sensor in general, and a synthetic aperture radar in particular, provides its own power through a transmitting antenna: this contributes to making radar a so-called 'all-weather' sensor by allowing acquisitions even at night. Moreover, since day and night acquisitions correspond to ascending and descending orbits, this also contributes to opposite side viewing capabilities so that hidden areas are greatly reduced.

Absolute location accuracy

Attitude (i.e. roll, yaw and pitch angles), which in the case of optical images is the major contributor of absolute location error, has no effect on location in a SAR image. Indeed, the computation of image point location in a SAR image requires only a slant range and a Doppler frequency shift, and these magnitudes do not depend on sensor attitude.

Sensitivity to relief

Radar image location is based on slant range measurements between the antenna and each terrain target: this is why terrain elevation has a direct effect on image location. This effect can be locally described as a proportional relationship between a height variation Δz and a slant range variation ΔR for a given incidence angle θ:

$$\Delta R = \frac{\Delta z}{\cos \theta},$$

where θ is the angle between the local vertical and the viewing direction.

At pixel scale, ΔR is a parallax that can be measured to derive an elevation: this is the basis for radargrammetry. At wavelength scale, ΔR can be estimated from a phase shift between two echoes: this is the basis for interferometry. Apart from this geometrical sensitivity, a radiometric sensitivity can be mentioned, since terrain orientation has an impact on radar return intensity: this is the basis for radarclinometry (or radar shape from shading).

Coherence

Radar signals are coherent, which means that they are determined in terms of amplitude and phase. The generation of a SAR image would not be possible otherwise. This characteristic makes radar sensors very sensitive to relief because of interferometric capabilities.

Atmospheric effects

The effects of the atmosphere and in particular the troposphere can be neglected at pixel scale, i.e. a radar image can be acquired and accurately located whatever the meteorological conditions. On the contrary, these effects cannot be neglected at wavelength scale, and they produce artefacts in interferometric products. For instance, a radar echo acquired under heavy rain conditions may have a phase delay of several times 2π even if the scene geometry has not changed.

1.6.2 Radargrammetry

Radargrammetry is an adaptation of the photogrammetric principles to the case of radar images. Indeed, it is based on parallax measurements between two images acquired from different viewpoints. However, radargrammetry cannot be carried out directly with photogrammetric tools, for two main reasons.

First, the SAR geometry is modelled by specific equations: this implies that analogue stereo plotters are not suitable, and that the first rigorous implementations could only be achieved with analytical plotters (Raggam and Leberl, 1984).

The second reason is that stereo viewing is difficult and uncomfortable with radar images, in particular in the case of rugged terrain. This is due to the fact that bright points and shadows suffer migration from one image to another due to the difference in illumination. However, stereo viewing capabilities can be acquired with practice, as relief perception is basically a psychological process (Toutin, 1997a).

The feasibility of radargrammetry was demonstrated for airborne radar data (Leberl *et al.*, 1987) as well as for spaceborne radar data (Leberl *et al.*, 1986). According to experiments made over different landscapes, it is now well known that this technique can provide 3D measurement with an accuracy of a few pixels, i.e. around 20 m with airborne SAR (Toutin, 1997b) and between 20 and 50 m with ERS (Toutin, 1996; Raggam *et al.*, 1993) or RADARSAT (Sylvander *et al.*, 1997; Marinelli *et al.*, 1997). As expected, the radargrammetric error tends to increase in case of steep relief.

1.6.3 Interferometry

SAR interferometry consists in deriving terrain elevation from the phase difference between two radar echoes acquired from very close antenna positions (Zebker and Goldstein, 1986; Massonnet and Rabaute, 1993). The two images can be obtained either from two antennas constrained to remain on parallel paths (this is generally the case for airborne systems) or from the same antenna during two passes. If these images are properly registered, the phase difference can be computed for every pixel and stored in an interferogram. Although the phase difference has a simple relationship with the slant range difference and therefore with elevation, relief cannot be derived so easily due to two severe limitations.

The first limitation is the dependence of radar phase on non-topographic factors, such as atmospheric variations (Goldstein, 1995; Kenyi and Raggam, 1996), land cover changes or sensor miscalibration (Massonnet and Vadon, 1995): these contributions to phase shift contaminate the interferogram and can be converted to elevation, contributing to the error of the output digital surface model. This is generally the case when the time interval between the acquisitions is too long, so that important changes have occured not only in the atmospheric refraction index but also in the surface roughness or electromagnetic properties.

The second limitation is the fact that the radar phase is not known in absolute terms but only modulo 2π, so that interferograms generally exhibit fringes that must be unwrapped.

The performance of phase unwrapping mainly depends of the *B/H* ratio, i.e. the ratio between the stereoscopic baseline and the flight height:

- when *B/H* is smaller, the fringes are wider: it becomes easier to unwrap them, but a given height gradient has less impact on the radar phase, which means that the interferogram is less sensitive to relief;
- when *B/H* is larger, the fringes become narrow and noisy: they are too sensitive to relief and they are more difficult to unwrap.

In the most usual configuration, interferometric images are acquired with a single antenna at different dates. The problem then is that the baseline cannot be predicted. Therefore, the accuracy of the method, which can be determined a posteriori, cannot be predicted before the acquisitions.

The most accurate results, which are close to the theoretical limits, are generally obtained with simultaneous dual antenna acquisitions. Interferometric processing based on the airborne TOPSAR system (Zebker et al., 1992) provided accuracies around 1 m in flat areas and 3 m in hilly areas (Madsen et al., 1995), and the CCRS dual-antenna SAR system provided a 10 m accuracy in mountainous and glacial areas (Mattar et al., 1994).

Single-antenna interferometry provides very heterogeneous results, due to the double influence of the spatial baseline (which cannot be predicted) and the time interval (which is generally too long) (Vachon et al., 1995). These influences vary according to landscape characteristics, and the accuracy depends on slope and land cover. Since these characteristics are often heterogeneous over 100 km side scenes, evaluating an interferometric accuracy over such a huge area is not very meaningful. Under suitable conditions (moderate relief, short time between acquisitions, no atmospheric variations), the error of interferometric products may be as small as 5 m with ERS data (Dupont et al., 1997) and 7 m with RADARSAT data (Mattar et al., 1998). As soon as ideal conditions are not fulfilled, errors often increase to several tens of metres.

During the generation of a radar interferogram, an associated product called the coherence image is usually computed as well, in order to display the correlation between the two complex echoes. This product provides useful information on the interferometric accuracy, because phase noise increases the output height error proportionally. However, correlation is also a very interesting thematic product, since correlation is an indicator of surface stability. For instance, rocky landscape and urban structures are very stable, so that they are generally mapped with high interferometric correlation, while forested areas and, above all, water, have very low correlation because the surface geometry within each pixel has permanent changes comparable to the wavelength (Zebker and Villasenor, 1992).

1.6.4 Radarclinometry (shape from shading)

While radargrammetry and interferometry are based on the principle of stereoscopy, and therefore on the geometry of the radar images, radarclinometry makes use of image intensity and can provide a DSM using a single radar image.

Radarclinometry is a particular case of shape from shading, insofar as it determines the absolute orientation of each terrain facet using its intensity in the radar image. In fact, this technique is a quantitative application of the visual perception of relief one has looking at a radar image.

The link between terrain orientation and radar image intensity is described by a backscattering model, in which the intensity is expressed as a function of the incidence angle (between the viewing direction and the perpendicular to the terrain surface) and the backscattering coefficient (similar to reflectance for optical imagery). Radarclinometry is basically an inversion of such a model. However, inverting a backscattering model is constrained by two major limitations:

- image intensity depends not only on the incidence angle but also on the backscattering coefficient and therefore on ground characteristics (land cover, roughness, moisture ...) which in most cases are not known;
- one particular value of the incidence angle corresponds to an infinity of possible terrain orientations, because surface azimuth or aspect also contributes to the incidence angle.

Due to these ambiguities, any radarclinometric algorithm must inject external constraints or hypotheses in order to make the inversion possible. To mention the most classical hypotheses, the first ambiguity is generally solved by assuming a homogeneous land cover (which implies that intensity variations are caused by slope variations only) and the second one by neglecting the slope component oriented along the track. A number of algorithms have been proposed to overcome these difficulties (Frankot and Chellapa, 1987; Thomas et al., 1989; Guindon, 1990; Paquerault and Maître, 1997).

Guindon (1990) obtained an elevation above 200 m from SEASAT data (with θ around 20°, i.e. a very deep incidence) over a mountainous area with steep slope: these are the worst conditions for radarclinometry. On the contrary, Paquerault and Maître (1997) obtained an error of the order of 20 m using RADARSAT over French Guyana, where the landscape is uniformly forested with gentle hills: this corresponds to ideal conditions for radarclinometry.

Finally, it should be noted that the elevation error is not very meaningful to evaluate the performance of radarclinometry, which is bascally a slope-mapping technique. The fact that standard elevation error is often used as a unique quality criterion for DSMs implies that very smooth surfaces are generally preferred, while radarclinometry is very sensitive to microrelief.

References

Dupont S., Nonin P., Renouard L. (1997) Production de MNT par interférométrie et radargrammétrie. *Bull. SFPT*, no. 148, pp. 97–104.

Frankot R., Chellapa R. (1987) Application of a shape-from-shading technique to synthetic aperture radar. *Proceedings IGARSS '97* (Ann Arbor), pp. 1323–1329.

Goldstein R. (1995) Atmospheric limitations to repeat-pass interferometry. *Geophysical Research Letters*, vol. 22, no. 18, pp. 2517–2520.

Guindon B. (1990) Development of a shape-from-shading technique for the extraction of topographic models from individual spaceborne SAR images. *IEEE Transaction on Geoscience and Remote Sensing*, vol. 28, no. 4, pp. 654–661.

Kenyi L., Raggam H. (1996) Atmospheric induced errors in interferometric DEM generation. *Proceedings of IGARSS '96 Symposium* (Lincoln), pp. 353–355.

Leberl F. (1990) *Radargrammetric image processing*. Artech House, Norwood, p. 613.

Leberl F., Domik G., Raggam H., Kobrick M. (1986) Radar stereomapping techniques and application to SIR-B images of Mt. Shasta. *IEEE Transaction on Geoscience and Remote Sensing*, vol. GE-24, no. 4, pp. 473–481.

Leberl F., Domik G., Mercer B. (1987) Methods and accuracy of operational digital image mapping with aircraft SAR. *Proceedings of the Annual Convention of ASPRS*, vol. 4, pp. 148–158.

Madsen S., Martin J., Zebker H. (1995) Analysis and evaluation of the NASA/JPL TOPSAR across-track interferometric SAR system. *IEEE Transactions on Geoscience and Remote Sensing*, vol. 33, no. 2, pp. 383–391.

Marinelli L., Toutin T., Dowman I. (1997) Génération de MNT par radargrammétrie. *Bull. SFPT*, no. 148, pp. 89–96.

Massonnet D., Rabaute T. (1993) Radar interferometry: limits and potential. *IEEE Transaction on Geoscience and Remote Sensing*, vol. 31, no. 2, pp. 455–464.

Massonnet D., Vadon H. (1995) ERS-1 internal clock drift measured by interferometry. *IEEE Transaction on Geoscience and Remote Sensing*, vol. 33, no. 2, pp. 401–408.

Mattar K., Gray L., Van der Kooij M., Farris-Manning P. (1994) Airborne interferometric SAR results from mountainous and glacial terrain. *Proceedings IGARSS'94 Symposium* (Pasadena), pp. 2388–2390.

Mattar K., Gray L., Geudtner D., Vachon P. (1998) Interferometry for DEM and terrain displacement: effects of inhomogeneous propagation Canadian Journal of Remote Sensing, vol. 25, no. 1, pp. 60–69.

Paquerault S., Maître H. (1997) La radarclinométrie. *Bull. SFPT*, no. 148, pp. 20–29.

Polidori L. (1991) Digital terrain models from radar images: a review. *Proceedings of the International Symposium on Radars and Lidars in Earth and Planetary Sciences* (Cannes), pp. 141–146.

Raggam H., Leberl F. (1984) SMART – a program for radar stereo mapping on the Kern DSR-1. *Proceedings of the Annual Convention of ASPRS*, pp. 765–773.

Raggam H., Almer A., Hummelbrunner W., Strobl D. (1993) Investigation of the stereoscopic potential of ERS-1 SAR data. *Proceedings of the 4th International Workshop on Image Rectification of Spaceborne Synthetic Aperture Radar* (Loipersdorf, May) pp. 81–87.

Sylvander S., Cousson D., Gigord P. (1997) Etude des performances géométriques des images RADARSAT. *Bull. SFPT*, no. 148, pp. 57–65.

Thomas J., Kober W., Leberl F. (1989) Multiple-image SAR shape from shading. *Proceedings IGARSS '89* (Vancouver, July), pp. 592–596.

Toutin Th. (1997a) Depth perception with remote sensing data. *Proceedings of the 17th Earsel Symposium on Future Trends in Remote Sensing* (Lyngby, June) pp. 401–409.

Toutin Th. (1997b) Accuracy assessment of stereo-extracted data from airborne SAR images. *International Journal for Remote Sensing*, vol. 18, no. 18, pp. 3693–3707.

Vachon P., Geudtner D., Gray L., Touzi R. (1995) ERS-1 synthetic aperture radar repeat-pass interferometry studies: implications for RADARSAT. *Journal Canadien de Télédétection*, vol. 21, no. 4, pp. 441–454.

Zebker H., Goldstein R. (1986) Topographic mapping from interferometric SAR observations. *Journal of Geophysical Research*, vol. 91, no. B5, pp. 4993–4999.

Zebker H., Villasenor J. (1992) Decorrelation in interferometric radar echoes. *IEEE Transaction on Geoscience and Remote Sensing*, vol. 30, no. 5, pp. 950–959.

Zebker H., Madsen S., Martin J., Wheeler K., Miller T., Lou Y., Alberti G., Vetrella S., Cucci A. (1992) The TOPSAR interferometric radar topographic mapping instrument. *IEEE Transaction on Geoscience and Remote Sensing*, vol. 30, no. 5, pp. 933–940.

1.7 USE OF AIRBORNE LASER RANGING SYSTEMS FOR THE DETERMINATION OF DSM

Michel Kasser

1.7.1 Technologies used, performances

Airborne laser telemetry (ALRS, an acronym that stands for airborne laser ranging system) is the normal term used in the 1970s for the airborne profile recorders (APR) that provided, in a continuous way, a very precise vertical distance from the airplane to the ground (for example Geodolite® of Spectra-Physics); the positioning of the plane was obtained then by photographic means that were not very precise in planimetry, the altimetric reference being directly the isobar surface at the level of the plane. These processes, fallen into obsolescence since the beginning of the 1980s because of insufficient precision of positioning, regained an interest when the GPS appeared as a very precise localization system, with data being processed after the flight. A large variety of equipment is available in 2000, including the laser rangefinder, its scanning device, an inertial platform and a GPS receiver, and of course the originality of every equipment resides in the proposed software packages.

The principle of functioning is as follows:

- The ALRS provides in a continuous way, at rates from 2 to 100 kHz, the time of propagation of a laser impulse given out by a powerful laser diode working in the near IR. This laser rangefinder is fixed to the plane as rigidly as possible. It may, in general, according to requirements, provide some different distances, for example that on the first

received echo (to measure objects over the ground, like an electric power line), or on the contrary on the last (to measure ground under vegetation), or again all echoes received for each shot (but then the quantity of data greatly increases). The precision of the rangefinder is in the order of some centimetres in the absolute, but in fact what matters is the interaction of the laser beam with the object targets, which is geometrically not very definite in most cases (e.g. high vegetation, building borders . . .). The beam is slightly diverging (typically 1 mrad), and even though the plane cannot fly high, considering the weak optic signal that is allowed not to create any ocular hazards, the analysis spot therefore has a diameter of several decimetres. A variant of this technology must be mentioned; it uses a continuous laser source modulated by a set of frequencies permitting ambiguity resolution (on the model of the old Geodolite® already quoted), but it is not widely used.
- An optico-mechanical scanner device allows the laser beam to be sent sequentially in different directions, which permits, with the advancement of the plane a scan of the ground, in a strip around the ground trace to be performed.
- In order to know the position of every laser shot in the space in spite of the unknown movements of the plane, the ALRS is coupled rigidly to an inertial unit platform. It provides information at a rate compatible with the movements of the plane, typically from 200 to 500 Hz. It must be initialized (some minutes or tens of minutes), so that its core components are in a stationary thermal regime. Its data errors are characterized by a bias proportional to the square of the time, which implies very frequent corrections.
- In order to correct the inertial unit that would drift quickly to excessive values without such external help, a precise GPS receiver is added to the whole. This receptor provides, in differential mode, the positions of its antenna as often as possible, generally between rates of 1 and 10 Hz, and the post-processing must provide precision compatible with the requirements of the study, which in turn implies naturally the localization of the reference GPS receiver in relation to the surveyed area.
- To complement this equipment, there is a PC unit to pre-process data and store the large quantities of observed data, to help with the navigation, etc.

The post-processing of measures is an intense operation. The GPS calculation, then the fusion of the GPS with inertial data, is an operation that may be automated to a large extent. Then it is necessary to process echoes, that is to say to filter non-applicable echoes, and this operation is not currently automated at all. For example to get a DTM under trees, it is necessary to identify 'by hand' which echoes probably come from the ground and which do not. Considering the number of echoes to treat, this operation, even with the help of the software, is very tedious. The impersonal and systematic character of the target selection (whatever the target, there is always an echo) is one of the fundamental features of this survey mode, and it may imply heavy data post-processing in certain configurations.

1.7.2 Comparisons with photogrammetry

Surveys by ALRS are significantly different from photogrammetric surveys, either using traditional digitized images or direct digital imagery. These surveys are in a semi-experimental period, in which the distributed and published information are not necessarily all applicable considering the commercial implications. ALRS surveys have therefore a low maturity compared to photogrammetry. One is nevertheless able to identify common features and weaknesses of these two techniques. An excellent survey on this topic is that by Baltsavias (1999), from which we take here certain elements. The main comparison between ALRS and photogrammetric surveys must include types of data process that are at a semi-experimental level in the two domains, which includes therefore, for example, developments in progress concerning automatic image matching and multi-stereoscopy.

Points that we will keep in this comparison are as follows:

- The DTM issued from ALRS can be available, if the type of work is suitable, in a very short time after the flight. For example, for the linear sites surveys without vegetation (urban surveys, power lines), the post process can be entirely automated, and the geometric data supplied in one or two hours. To the contrary, the supplying of DTM by automatic image matching requires processes that are not yet completely automatic, and are always quite long.
- The equipment used in ALRS is sophisticated and expensive, and some unit components may experience some export limitations according to their countries of origin (USA for example). The maintenance when in use is therefore still not very easy. We are in the presence of technologies of which some are very recent, evidently in contrast with photogrammetry where devices of aerial image acquisition are industrialized at a very high level, and are of a remarkable reliability.
- The installation of the ALRS equipment on board a photogrammetric plane or a helicopter is a delicate phase that requires, for example, aiming toward the ground, which is often not an easy situation. Besides, the ALRS equipment requires a GPS antenna (delicate to install on certain helicopters) and consumes electric power that is not always available on the current planes. The geometric link between different

subsets (GPS antenna, inertial power station, ALRS) is a delicate metrology operation, but fortunately it can be checked by the airborne measures themselves if the software of calculation foresees it.
- The survey by ALRS is an active technology that doesn't require, in any aspect, lighting by the sun. There are no limitations therefore to timetables of flight, other than those dictated by the security of the aerial navigation. Flights are necessarily performed at low altitude, therefore in zones with little problem of flight authorization (out of the commercial air space), but which on the contrary are more and more difficult to get over cities. But this possibility of working in the morning, in the evening or in winter doesn't present the same comparative advantage now that one may use digital cameras, whose light sensitivity is such that one can acquire excellent pictures without difficulty with a low level of ambient light (low sun, flight under clouds). As long as the question is to provide some DTM, such flights under weak lighting are quite acceptable and competitive with the ALRS, but the current problem is that one very often asks *also* for an aerial image acquisition to serve as a basis for photo-interpretation, and even the production of orthophotographies, and this requires good sun lighting on the other hand. However, the ALRS generally cannot provide a picture, in complement of the DTM: it is a tool to provide DTM and nothing else. It doesn't allow one to discern in a simple way a house from a tree, for example, which limits its field of application quite a lot. And if one adds a camera in the plane, considering the height of flight it leads to a very significant image quantity, so that one would not know how to deal with the present means in order to make a mosaic or an orthophotography, even though all necessary data are available.
- Interactions of the ALRS with vegetation are interesting and numerous. To the list of problems, let us note that echoes obtained in the high vegetation are very difficult to assign to a specific layer, that is one works with the first or the last echo detected. On the other hand, in forests the ALRS is the *only* technique that can provide, in practically all types of forests (even tropical), a DTM of the ground through leaves and branches. It presents a major interest therefore for planning in equatorial zones, where the hydrology is dominant and cannot be correctly processed without such data. (See Figure 1.7.1.)
- An interesting feature of the ALRS resides in its capacity to measure some echoes even on very small objects. The most spectacular example is that of high-tension power line surveys. It would be impossible in photogrammetry to survey lines, not because one doesn't see them (with a digital camera one sees easily even thin wires thanks to the considerable dynamics of the picture), but because there is no means to aim at the cable when the stereoscopic basis is parallel to the line,

Airborne laser ranging systems for DSM 57

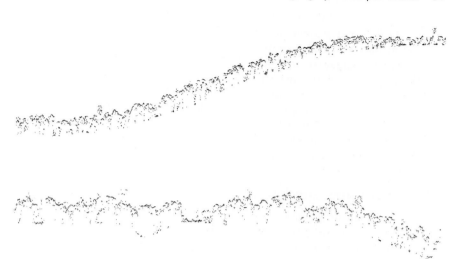

Figure 1.7.1 Two raw profiles obtained by the ALTOA® ALRS in French Guyana (tropical vegetation around 40 m high). One notes that the shots ranging to the ground describe very finely the topography, and that in such circumstances a photogrammetric survey using the tops of the canopy and subtracting a constant value would (1) ignore important morphologic features such as trenches and (2) induce large errors as the height of the canopy is far from constant.

which is necessarily the case for a linear survey of this type. In high-tension power line surveys by ALRS, it is possible to survey almost in an entirely automated way lines and the zones of land and vegetation that are close, which permits the identification on time of the interventions of lopping that could be necessary.

- In most other domains, there are few differences between obtaining a DTM by ALRS or by photogrammetry. Some authors consider that in urban situations the ALRS presents significant advantages, but one can note many counter-examples, and one may specify that the multi-image matching (automatic correlation with more than two images) on images of large dynamics in the city also provides excellent results, with advantages and drawbacks that are different.
- We must note that error models are not at all the same, and that neither of the two techniques really surpasses the other. For example punctual mistakes are rare in photogrammetry, and numerous in ALRS (electric lines, birds, reflection on planes of water, etc.). The precision of measurement of a building may be also analysed in different ways in the two cases: the automatic matching will easily find the edges but will be in difficulty because of the hidden parts, the ALRS won't describe the edges correctly because of the width of the analysis spot.

For example, with regard to measures of pure altimetry, one cannot establish a net benefit in terms of precision, the ALRS being locally exact, but not being capable of identifying ditches or very important, although small, streams on such sites. The data provided by ALRS or by automatic image matching in photogrammetry are not originally structured. Numerous studies are in progress to structure them in an automatic way at least for the simple cases (e.g. buildings). And in this domain photogrammetric methods use extensively the possibilities of automatic analysis of every image, whereas the ALRS data are much more difficult to analyse.

1.7.3 Perspectives, conclusions

One is therefore in the presence of a new technology, whose optimal use in cases probably must be deepened again, but that one has nevertheless to take into account from now on for data production (Ackermann, 1999). The ALRS must be considered as a complement to photogrammetry, especially suitable in some cases:

- surveys under vegetation;
- linear sites DTM and very fast supplying of narrow zones;
- high-voltage power line surveys;
- bare surface DTM.

There are a great many studies in progress to improve the fruitfulness of this technique, and costs will probably soon become clearer, considering their present instability.

References

Ackerman F. (1999) Airborne laser scanning – present status and future expectations. *ISPRS Journal of Photogrammetry and Remote Sensing* 54, pp. 64–67.
Baltsavias P.E. (1999) A comparison between photogrammetry and laser scanning. *ISPRS Journal of Photogrammetry and Remote Sensing* 54, pp. 83–94.

1.8 USE OF SCANNERS FOR THE DIGITIZATION OF AERIAL PICTURES

Michel Kasser

1.8.1 Introduction

The present need for digital pictures destined for photogrammetric processing is probably not going to stop growing in the next few years.

It was at first bound to developments in the digital orthophotography market, from the 1980s. This development is far from finished, especially because users are now able to handle without difficulty large digital data quantities on low-cost personal computers, and particularly pictures. So in a lot of urban GIS (geographic information systems) where data must frequently be updated, it is current practice to work on a basic layer formed by an orthophotography, that is not costly and therefore can be often changed. The vector layers are then locally updated, from the orthophotography, directly by the user himself. This evidently creates an important market for digital aerial pictures. In addition, since the beginning of the 1990s many software developments have been made, intended to do all the photogrammetric restitution directly on PCs, transformed thus in digital photogrammetric workstations (DPW, described in §2.5) of lower and lower cost. Users of such DPW are no longer only societies of photogrammetry, but are more and more technical users coming from sectors quite external to photogrammetry, who themselves do the restitution of objects that interest them. The diffusion of the DPW also creates a demand for digital pictures, demand that is likely to increase for a few years.

This need for digital pictures is currently satisfied in two ways: by digitization of traditional silver halide pictures, and by the more and more frequent use of digital aerial cameras (see §1.5). Pictures obtained by these cameras have much better performances (radiometric and geometric as well), but the cameras' rarity has made it necessary for several years to use scanners to digitise the new images acquired by traditional aerial cameras (which are in general very reliable and of quite long life), and these devices will remain necessary for a long time to allow the use of old pictures.

We are going to analyse the techniques used by these scanners in order to understand their shortcomings and their qualities. These materials themselves are evidently also in regular evolution, but this evolution is notably slow and the technologies used seem steady enough, probably because there were no innovations in the domain of sensors applicable to these devices for some years. We describe here the situation in the year 2000: there are two classes of scanners, adapted to aerial pictures 24×24 cm. There are those that have specifically been developed for photogrammetry that are very precise, and thus very costly considering their low diffusion (Kölbl, 1999), and to the contrary these are scanners of A3 format, less precise but for a large public, much cheaper, and that can also be used.

The specification of these scanners is as follows: to allow the processing of aerial pictures 24×24 cm, pictures on paper or on film, an analysis pixel smaller than or equal to 30 µm, in colour or in black and white, with a good geometric precision. The possibility of processing some original roll film directly is also an important specification for a part of the market.

1.8.2 Technology of scanners

1.8.2.1 Sensors

Sensors used by off-the-shelf devices in 2000 are all based on charge-coupled devices (CCDs, already seen in §1.5), being combined with an in-a-plane analysis of images. Photomultipliers were previously used with an installation of images on a drum, which was not as easy, but this generation of equipment is no longer available. These CCD are used in three ways: in linear CCD, in matrixes, or possibly in TDI linear CCD. Let's look at these sensors in detail:

- The linear CCD are used in order to be able to scan at once with a set of 10,000 or 12,000 detectors, either on only one linear CCD, or while joining several shorter linear CCD. To get an analysis of colour, one uses three of these parallel linear CCD juxtaposed, each equipped with one of the three RGB filters. The group formed by these linear CCD and the necessary optics is then displaced according to only one axis in relation to the picture to be digitized, either it moves before the stationary image, or the inverse. The scan of a whole picture requires several successive passes, which therefore have to be perfectly connected.
- CCD matrixes used have the order of $2,000 \times 2,000$ pixels. There are much larger matrixes, but given their high cost they are not a good choice as their use would not appreciably accelerate the process. The detector and its optic move in a sequential way according to two measurements in front of the image to be processed, in order to cover all the useful surface (it is necessary according to the size of the chosen digitization pixel to cover up to $30,000 \times 30,000$ points of analysis), and there again the mechanical displacement must permit an excellent connection between the individual pictures that will make up the whole image.
- The linear CCD functioning according in TDI mode (TDI for 'time delay integration', an inauspicious enough acronym because it explains nothing). One exploits an oblong matrix (for example $2,048 \times 96$ pixels), and one uses it like a linear CCD (here of 2,048 pixels). While making it advance regularly during the digitization as if it were a normal linear CCD, one transfers charges produced from a line to the following one (here 96 times), precisely at the same speed as the displacement of the picture in the plane of the CCD. Thus a whole set of charges recovered at the exit has been generated by only one point of the analysed picture. So these charges have been produced by many successive detectors (here 96), which homogenises the responses of all detectors to a remarkable degree. It is similar to the way of working that is described in §1.5 to perform the forward motion compensation on matrix airborne cameras. One gets therefore a very regular response from all detectors during the scan.

Scanners for aerial image digitization 61

These devices present different advantages. For example, the time of integration of a pixel facing the analysed photograph will be much longer for a matrix than for a simple linear CCD, which will permit one to use a less powerful lighting to illuminate the film, and will prevent undesirable thermal effects due to the powerful lamps. The analysis of the colour by three neighbouring linear CCD equipped with filters is simpler to achieve than with a matrix, for which it is necessary to commute a set of filters sequentially before every acquisition of an elementary zone. On the other hand, if filters are put down directly on linear CCD, which is generally the case, it is not possible to modify them, and there is no means to regulate differently the lighting of each channel, which is possible with a matrix system. Obviously it is necessary that the focusing of the optics be carefully tuned, and with linear CCD systems it must be the same for the three linear CCDs, which is not easy considering differences of wavelengths and therefore of diffraction effects, which are very appreciable for extremely small pixel sizes.

Let us note that irregularities of detector radiometric response may obviously be corrected by a preliminary calibration, but that those that subsist (dust particles, calibration errors) will create periodic and annoying artefacts:

- for linear CCD, some parallel strips will be displayed on the whole document, among which one will also note those due to the successive passes of linear CCDs;
- for matrixes it will be due to the repetition of errors according to a regular paving, and of radiometric discontinuities between successive positions of the matrix. These shortcomings will be generally far less spectacular than for linear CCD.

Besides, behind every CCD there are special electronics (voltage-current amplifier, digital to analogue converter) that must be optimized so that the reading noise of the CCD is as low as possible, with a digital sampling ranging from 8 to 12 bits. But this very high quality sampling is not so necessary considering the noise of the photographic process itself, which does not justify more than 6 useful bits. One will find normally all classic defects of the CCD (echoes, blooming . . .) that are described in the corresponding technical literature (Kodak, Dalsa, Phillips, Toshiba, Thomson . . .).

1.8.2.2 Mechanical conception

To displace an optic as a whole and to reposition it elsewhere within some microns is not a simple operation. It is also one of the major technological difficulties of scanners. That one uses linear CCD or matrixes doesn't create major differences in this respect. It is therefore mainly problems of mechanics, that are generally solved using mechanisms having some extremely reduced play, achieved in materials only giving a weak coefficient of differential thermal dilation between the various sub-assemblies: one does require from this mechanics only that it be extremely reproducible in its errors. One proceeds to a very tidy calibration of the whole while observing a regular and known pattern with a very high precision (patterns on glass plate, known to about 1 μm). One then deduces a set of correction parameters to apply to the different elementary pictures assembly in order to reach a precision compatible with the specifications (generally, some μm). But it is necessary to be very prudent with the ageing of these structures, because of unexpected effects of dust on movements, or because of any play that may develop and compromise the validity of tables of correction parameters as well.

1.8.3 Size of analysis pixel

The available pixel sizes are very variable, starting at 4 μm on the best resolution devices up to more than 300 μm when one achieves a coarse sampling. Generally these large pixel sizes are reached by regroupings of data obtained on the smallest possible analysis pixel. This generalized use of a very small pixel and its regrouping by software is also used to rectify the geometry of under-pictures between them, which without re-samplings are still very expensive in time calculation and in memory space.

Recommended sizes of pixel evidently depend on the quality of the image to be digitized and on the photogrammetric work to be performed. In §1.9, where Christian Thom studies problems bound to the signal/noise ratio of photographic emulsion, one will see that the smaller the pixel, the more the signal/noise ratio deteriorates, and this well beyond the possible defects of the scanner. A size of 30 μm is satisfactory in most of the studies of aerotriangulation type, DTM, orthophotography, etc. If one uses smaller sizes (e.g. 15 μm), gains in precision are often modest (Baltsavias, 1998) but at a high cost in terms of computer complications.

1.8.4 Radiometric quality, measure of colours

Considering the quite mediocre radiometric quality of photographs (with a fidelity to the original object that is weak considering the chemical process used), it is certain that there are not many conceptual difficulties in achieving an analysis of images that practically does not bring any data deterioration. In particular the restitution of colours, in such conditions, looks more to the operator's artistic sense (to provide a document satisfying the customer's eye) than to a rigorous respect of physical laws. The available devices offer a large palette of tools in order to best balance the RGB components. Let us note, however, that some mechanical defects can drive to a small geometric colour component shift, which will be able to generate some local artefacts especially visible on contrasted objects with straight edges, linear or of a size close to the pixel size.

1.8.5 Large A3 scanners

Scanners to the format A3 are available on the market, which are therefore extensively capable of digitizing a whole aerial photograph, and intended for the technical public in general. They are incomparably cheaper than the specialized devices for photogrammetrists (around 50 times less) and it is interesting to note their possibilities of use. Those available have to the date of writing, sizes of active pixel from 30 to 40 µm, using a triple linear CCD with RGB filters, which permits a digitization in only one pass and therefore a good geometric homogeneity, some geometric errors ranging to 2 pixels that generally create an affine distortion. It is to be noted that such distortions are also typical of films, because of processes that they undergo at the time of the development and drying, and that these errors are systematically modelled by one additional unknown in phases of aerotriangulation. It is therefore possible to use such scanners for certain photogrammetric studies not requiring the maximal precision, and a fine assessment of their limits and the temporal stabilities of error models is to be done for each of them.

References

Kölbl O. (1999) Reproduction of colour and of picture sharpness with photogrammetric scanners, findings of OEEPE tea scanner test, *Proceedings of tea OEEPE Workshop on Automation in Digital Photogrammetric Production*, Marne-La-Vallée 22–24 June, pp. 135–150.

Baltsavias E.P. (1998) Photogrammetric film scanners, *GIM*, vol. 12, no. 7, July, pp. 55–61.

1.9 RELATIONS BETWEEN RADIOMETRIC AND GEOMETRIC PRECISION IN DIGITAL IMAGERY

Christian Thom

1.9.1 Introduction

One of the first problems which the digital picture user is confronted with is the size of the pixel that he is going to use for his scan. Intuitively, he feels sure that, the smaller the step of the scan, the better will be the precision, at least relatively, of the result, at the cost unfortunately of a greater data volume. But generally he does not know that this better resolution will also be paid for, as we shall see, in terms of radiometric quality. This loss of radiometric precision also has an indirect but certain consequence on the geometric precision. One will understand this, according to the principle that 'who can do more, can do less' (one can always gather some small pixels to build bigger ones ...). Nevertheless, one does not risk getting a poorer result, paradoxically with small pixels than with big ones. But as the size finally chosen is still the result of a compromise between the geometric precision and the cost/volume of data/time processing, it is clear that one must take into account the effect of the radiometry on geometry to find the best compromise.

What is true for digitized pictures is true for digital pictures, that is to say those provided by digital cameras, and this is because of two points. First, a better geometric resolution itself has a cost, because it intervenes directly, and fairly linearly, on the cost of the image acquisition: it is not a question of a simple adjustment of the scanner, but is rather like the choice of the scale in the classic case. Second, digital picture radiometric quality is much better than that of the digitized pictures (under usual conditions of digitization). One may easily understand that the aforementioned effect is more crucial in this case.

A poor radiometry has obvious effects on other aspects than the geometric precision, for example aesthetics and the interpretability of the picture; they go far beyond the scope of this chapter because they are too subjective to be modelled in a rigorous manner. Yet, they are not completely different because the geometric precision that one can reach in a picture certainly has an influence on its interpretability, and maybe on its aesthetics.

1.9.2 Radiometric precision

This notion is relatively simple. Radiometry is the measure of the energies given out by the objects of a scene. Its precision is therefore very definite, as for any measure. However, in the context of this survey, the radiometry in itself is still not accessible to us. Indeed, if the digital picture sensors are almost always radiometric, this is not the case with the digitized film, where the digital data of the picture are functions of the radiometry, monotonic-growing and limited, but not well known, and that, besides, depend on the conditions of development of the emulsion. Fortunately, it will only have little impact on our analysis, because we try to value the precision of localization of visible detail in the images, and this precision will always be a function of the ratio between the noise and the contrast of the detail. This ratio may be valued more or less directly in the image.

If the notion is simple, its evaluation is not, especially in the case of the digitized pictures. If models of noise are well known for the digital sensors, they are not in the case of digitization, where the noise of the scanner, noises bound to the emulsion and its process superimpose themselves. Here, we will use data that we acquired from experiments (1), and from measures on the digital camera of IGN-F (2). However, it is necessary to investigate the origin of these noises, because some aspects have an impact on our topics.

1.9.2.1 Digital camera

The noise model here is quite simple. The sources of noises are:

1. the reading noise of the sensor, due to the electronics of sampling, essentially Gaussian;
2. the photon noise, due to the corpuscular nature of light, according to a law of Poisson;
3. differences of pixel sensitivity (may be corrected by calibration);
4. differences of dark current (may be corrected also by calibration).

In general, it is the photon noise that dominates. It grows according to the square root of the signal, and its value depends on the total number of photons that every pixel can contain. For example, in the case of the KODAK sensor used at IGN-F, this number is 85,000, and the signal/noise ratio at full capacity is 300.

1.9.2.2 Digitized pictures

Sources of noises are here more numerous and not very well known:

1. noise coming from the sampling unit, which depends on the equipment used, the size of pixel, the level of grey obtained, etc.;
2. noise coming from the emulsion.

Let us look at this second point in more detail. One knows that the photographic process is based on detection of light by ions of silver, then to an amplification of the signal by the developer, that makes every nucleus created by a photon detected in a grain of silver of a certain size grow (some microns). The film is then analysed by the scanner that values its optic density, that is to say the proportion of light that passes between grains of silver. One immediately sees numerous consequences:

- The response of the film is not linear. Indeed, the photochemical process of grain production is not linear, because several photons are necessary so that a grain can be created, producing low sensitivity of the film to low illuminations. But even though one may suppose this process as linear, when the density of silver grains grows, they make 'shades', when a grain overlaps another. The marginal sensitivity for a proportion of R grain will thus be $1 - R$. This gives us a function according to the illumination i of $R(i) = 1 - \exp(-ki)$. The optic density, which is proportional to the logarithm of $1 - R$, is therefore also more or less proportional to the illumination of the emulsion.
- If one considers the middle of the dynamic range of the film ($R = 1/2$), for a size S of the pixel analysis and a characteristic size of grain of s, if i expresses the number of photons detected, one has:

Surface of the pixel: S^2
Surface of the grain: s^2
For small values of i one notes that $k = s^2/S^2$
For $R = \frac{1}{2}$, $i = \log(1/R)/k = \log(2) \, S^2/s^2$
The signal to noise ratio is

$$r = \sqrt{i} = \sqrt{(\log 2)} * S/s. \tag{1.33}$$

Let us take, for example, the following values: $S = 20$ μm, $s = 1$ μm: one finds $r = 16.6$. Recall that under similar conditions, a sensor directly digital may benefit from a signal/noise ratio of about 150. This gives an indication of the problem ...

If we consider the values obtained, it seems justifiable to neglect in what follows the scanner's own noise in as much as we don't know it, and as it depends on the type of device used.

One also notices in this formula that r is proportional to the dimension of the analysis spot. In general the spot of analysis on a scanner is bigger than the step of sampling to avoid any aliasing, which slightly improves the previous values, at the cost of a poorer MTF (modulation transfer function) of course.

Finally, this formula gives us an idea of the ratio between the quantum efficiency of the film and of the CCD sensors. Indeed, our experience shows us that conditions of image acquisition (exposure time, relative aperture) in the two systems are practically the same. One sees that on a pixel of 20 μm, the film detects 280 photons, whereas on a pixel of 9 μm, the CCD detects about 20,000 of them, which corresponds to a ratio of 350 in sensitivity ...

These very simple equations explain why the CCD sensors can be smaller than the usual focal planes of classical cameras, and that their pixels are themselves smaller, but that they retain nevertheless an enormous advantage in terms of radiometric quality. The entire problem is to exploit it to the fullest, to compensate their relative apparent lack of resolution.

1.9.3 The geometric accuracy

We will concern ourselves here in the geometric accuracy in the picture, and not in the accuracy of objects in the scene. This means that we will make abstraction of all problems of systematism bound to the image acquisition, even if they may be important (e.g. the aerial cameras based on linear CCD sensors), where the poor knowledge of parameters of external orientation of the sensor leads for every line of the picture to a poor precision of pixel localization.

One can distinguish two different ideas in this domain. First of all, the precision of positioning of a detail in the picture, this detail being able to be punctual, linear, or area, this one going back to the linear case in general because we position its limits most of the time. In the same way the punctual is often the intersection of two linear details (corners of building, for example), although there are also some merely punctual details. The fact that the detail is distributed in general on several pixels obviously helps its localization, because one can have several evaluations of the measured position, and therefore take the mean value.

Then, there is the more delicate problem of the separation of two nearby details, or resolution, which affects the interpretability of the picture more than the precision itself, but which we mention here because it is more demanding in terms of pixel size. Indeed, it is clear that even with pictures of very good quality it is impossible to separate details less than two pixels apart, whereas in terms of localization precision, one can without difficulty, in the same conditions, aim at a few tenths of a pixel. If the size of the pixel is evidently an element determinant of this precision, it is not the only one. The resolution of the optics is another. All situations can present themselves between two extremes:

- fuzzy pictures over-sampled with pixels which are too small;
- high-resolution pictures under-sampled with pixels which are too large.

Both cases can be found for example with a scanner either poorly set up or not tuned, and the second with digital cameras not having adequate optics. These situations evidently include the ideal case of pictures sampled while respecting the limit of Shannon, but it is very rare, because the characteristics of optics are not constant in their whole field of view, and therefore a regular sampling does not permit the criteria of Shannon to be respected everywhere.

One is confronted therefore, in general, often with no ideal and sometimes with unknown situations. We will examine some simple typical cases, in general at the extreme ranges of the possible, the intermediate situations being always more complicated.

1.9.3.1 Precision of localization

The case of under-sampled pictures

With regard to the punctual details, the precision is easy to determine: there is no means to determine the position of the detail inside its pixel (see Figure 1.9.1). We deduce a RMS precision in x and y of 0.29 pixel, independent of radiometric precision. In the case of a linear detail, the precision that one can get depends on the a priori knowledge one has on the nature of the detail (is it straight?), and of its orientation in relation to the axes of the picture. The worst case is evidently a detail parallel to one of the axes of the picture, and having the same precision of 0.29 RMS pixel. On the other hand, for a differently oriented detail, the distribution of the energy in the different pixels crossed by the detail permits the precision to be improved, which depends then on the characteristic dimension of the detail by which one can consider it as straight. Since one uses the value of pixels crossed, radiometric precision will have an importance, but this is difficult to evaluate. With regard to area details, the problem is to position the edge of the surface. This situation, fortunately more frequent than the previous (fields, roads, buildings, etc.), is more favourable, because sub-pixel position of the edge is function of the quantity of energy received by the pixel containing the edge. If we simplify the problem taking an edge parallel to the y axis, of c contrast, one understands that its sub-pixel position in x can be calculated by: $dx = dg/c$, dg being the difference of radiometry in relation to the pixel of reference. The precision on x is therefore directly bound to the radiometric noise. To give an idea, with a contrast of 1/10 normalized to the value of saturation of the sensor, one gets in the case of the film digitized at 20 μm a noise with a rms of 0.6 pixel, which means that no profit is possible, and in the case of the digital sensor $1/15 = 0.07$ pixel, which is substantial. Let us push a little farther the problem of the film. Indeed, why use smaller pixels to the digitization?

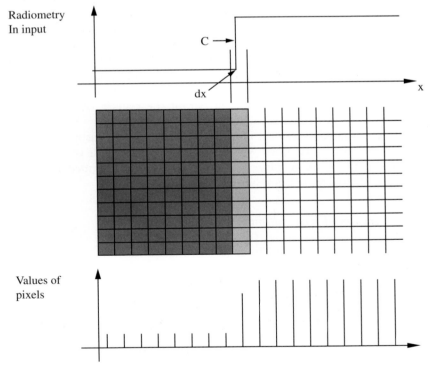

Figure 1.9.1 Limit of zone, under-sampled situation.

The noise grows as $1/S$, and therefore if one is only interested in a pixel, the precision will be the same, whatever the value of S. Fortunately, it is necessary to take into account the fact that for a unit of length of the detail, one can take the mean value of the position of the detail as many times that it contains pixels. One will have therefore a precision proportional to $s.(S/L)^{1/2}/c$, where L is the usable dimension of the detail. In fact one cannot decrease the size of the pixel indefinitely without reaching the limit of resolution of the optics, and therefore moving to the next case.

The case of over-sampled situations

We study here the case where any transition of radiometry is progressive. All detail is represented therefore on several pixels. There, all the previous cases are reduced to the case of one edge. Indeed, a linear detail is described by two successive edges, their characteristic dimension being, however, the half of one edge (the transition of an edge is in fact the integral of that of a linear detail, i.e. the point spreading function (PSF), projected on the axis perpendicular to the detail). We will then look at the localization of a zone of constant gradient of width D and useful length L. The D size is bound to the width of the PSF of the optics. One can consider that it is the half-maximum width of this one in the case of the side of an area detail, and of one half of this one for a linear detail. Our model here is very simple, but it will be nevertheless sufficient to analyse the problem in a qualitative manner (see Figures 1.9.2 and 1.9.3).

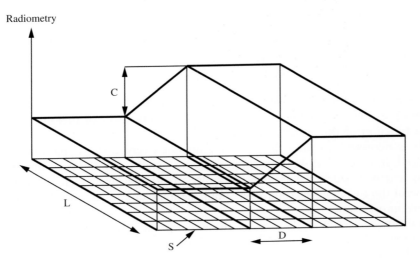

Figure 1.9.2 Limit of zone, over-sampled case.

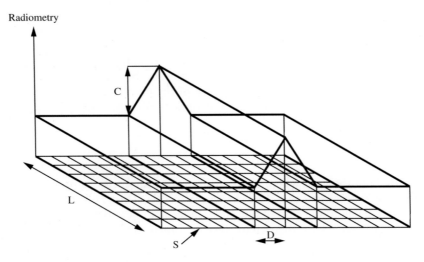

Figure 1.9.3 Linear detail, over-sampled case.

The precise position (sub-pixel) in x is given by the mean value of the zone where the gradient is constant, divided by this gradient:

$$Dx = \frac{\sum g}{LD} S^2 \frac{D}{C} = \frac{\sum g}{L} \frac{S^2}{C}. \tag{1.34}$$

The noise on this value is characterized by:

$$\sigma_x = \sigma_g \frac{S}{\sqrt{LD}} \frac{D}{C} = \sigma_g \frac{S}{C} \sqrt{\frac{D}{L}}. \tag{1.35}$$

Here again, in the case of a digitized image, the size of pixel S also intervenes into the denominator, and therefore is independent of this one, but is a function of the grain size.

It is necessary to note that one should not overuse the L parameter. Indeed, it is rare that one can suppose that a detail is linear on an important length. Besides, we assume we know its orientation, which is not in general the case, and introduces a supplementary unknown, and therefore more imprecision.

Let us recall that in the case of the linear detail, the assessment of the position can also be done with the other side of the detail, and that D is one half of the previous case. Thus the benefit is in relation to the edge length, a factor 2 at the end.

General case

In fact, in practice one is most often confronted with intermediate situations, where the PSF of the optics is the order of the pixel, or of a pixel fraction, and is not known with precision, or varies one point to another, etc. The methods of precise positioning described may give in this case mediocre results with residual bias. This is due to the fact that the PSF is itself poorly sampled. Yet these methods always bring an improvement in relation to the situation seen in the case of under-sampled pictures whose results can be considered therefore to contain major errors.

Conclusion

One can extricate several findings from what we have just seen:

- to work with well-sampled pictures is important if one wants to fully benefit from this kind of technique;
- however, to position the edge of area details, a poor sampling is sufficient;
- the use of this kind of technique on the digitized images is useless, except for very contrasted details, the noise in this type of pictures being far too dominant.

1.9.3.2 Spatial resolution

Thin detail detection

Contrary to what one commonly believes, it is not impossible to see in a picture details smaller than the size of the pixel. Just note that it is sufficient that the contrast of the detail, multiplied by the ratio of the surface occupied by the detail to that of the pixel, is in correct relation with the radiometric noise.

An example of this property is given in Figures 1.9.4 and 1.9.5. These are pictures taken on the same zone, a pond over which passes several high-tension power lines. The first image is an excerpt of a picture coming from one digital camera of IGN-F. The second image is a digitized image. Sizes of ground pixel are similar, around 75 cm. On these two excerpts, one distinguishes cables of the thickest line (probably doubled), but the finest line is visible only on the picture from the digital camera. The section of cables is 3 cm, that is to say of the order of the 1/20th of the pixel size! One sees that here also, a good signal to noise ratio in pictures allows one to compensate a poor geometric resolution.

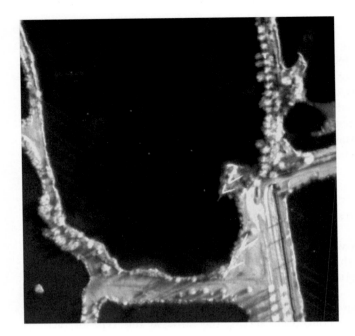

Figure 1.9.4 IGN-F digital camera.

Figure 1.9.5 Digitized image.

Resolving power

It seems at first that radiometric quality is not able in any way to compensate for a poor geometric resolution. Indeed, two details situated in the same pixel will never be able to be separated. This proposition is obvious, but it is necessary to state it however.

One can view the problem in the opposite direction, that is to say to interrogate oneself on the real resolving power of pictures whose radiometric quality is insufficient. Traditionally one is supposed to discern two details (choose for example bright ones) if they are separated in the picture by dark pixels. It is clear that in the case of details poorly contrasted, and in pictures of poor quality, the bright pixels because of the presence of details will have a certain chance of being darkened by the noise, and the dark pixel that separates them will have a possibility, to the contrary, to be brighter. The inverse situation may happen, where an object will be split falsely. One sees that one rejoins here the problem of segmentation, since one poses the question: 'Does this pixel belong to this object?'

Let's study a concrete case, for example: the segmentation of a supposed homogeneous zone. We deliberately choose a simple algorithm, where a step on radiometry determines the assignment to one zone or to another. If the contrast between the two zones is c, the step will be placed to $c/2$ (one supposes the noise to be independent of radiometry). What is, in these conditions, the probability that a pixel is poorly classified? It is a simple problem of statistics, bound to the interval of confidence. The success of the segmentation process and its precision will obviously depend on this parameter, as a non-trivial function of the algorithm used, and a priori knowledge that one has from structures in the picture (straight edges for example).

One can see in Figures 1.9.6 and 1.9.7 the same zone in two simultaneous image acquisitions, one with a digital camera of IGN-F, the other with a traditional camera. If one pays attention to the ridges of roofs, one notices that they are clearly delimited in the digital picture, whereas some edges are indiscernible on the digitized picture. This loss of visibility will evidently have consequences on the precision of its restitution, notably with automatic means. It is unfortunately impossible to quantify this here, because it evidently depends on the algorithms used.

In Figures 1.9.8 and 1.9.9 one can see the result of the filter of extraction of contour from Adobe Photoshop for example. It clearly appears that the extraction of roof edges can be performed more easily and more precisely on the digital picture. Note the side of roads in the shady regions.

To illustrate even better the influence of radiometric quality, and its interactions with geometry, see Figures 1.9.10 and 1.9.11: the same picture is presented here, but after it has been under-sampled by a

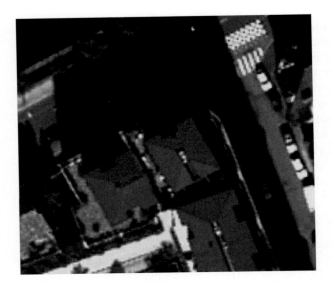

Figure 1.9.6 IGN-F digital camera.

Figure 1.9.7 Scanned image.

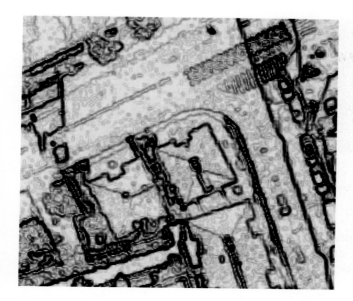

Figure 1.9.8 IGN-F digital camera, contours.

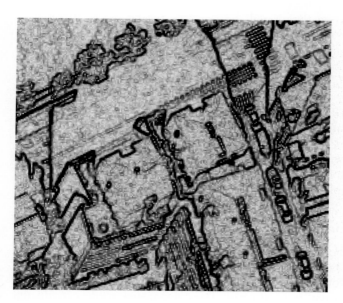

Figure 1.9.9 Scanned image, contours.

Figure 1.9.10
Under-sampled scanned picture.

Figure 1.9.11
Contours from Figure 1.9.10.

factor 2. This operation, as it averages pixels, decreases the noise, and one sees that the process of contour extraction now works better, but its result will be evidently less precise, being obtained with a pixel twice as large.

We didn't mention here the question of the sampling technology, but it has an influence of course. When the PSF is larger than the pixel, the contrast between two nearby details tends to soften, and therefore the restitution of the two details will be sensitive to the noise. When it is smaller, one doesn't note an appreciable effect.

Conclusion

It is clear that for what concerns the geometric resolution, radiometric quality brings much with regard to the detection of details. It has an influence on the separation in the case of weakly contrasted details, that is to say whose contrast is the same level as the noise present in the picture. It is necessary to mention that these details are frequent in urban zones, materials of construction often being of similar albedo (different types of coating, for example), and the frequent shade zones.

1.9.4 General conclusion

The impact of radiometric quality on the geometric quality is still not easy to evaluate. The problem is complicated by the fact that it depends on algorithms used for the restitution, and on the adequacy of the picture sampling. Yet, it is real in most cases. The comparative studies conducted on the digital camera of the IGN-F and on digitized pictures, to scales ranging from 1/15,000th to 1/30,000th, clearly show that the very large available dynamics of the CCD camera allows one to reach the same uses as the digitized silver halide picture, this in spite of a pixel of a size about twice as large. But this coefficient of 2 is only valid in this range of scales, and other studies should specify its value for smaller pixels: this coefficient doesn't depend indeed only on radiometric quality, but also on the type of object being surveyed.

2 Techniques for plotting digital images

INTRODUCTION

This section deals with the ways of extracting the 3D data from the images. Some considerations, here also, are not really specific to the use of digital images, but they must nevertheless be examined within the scope of digital photogrammetry, as in the case of aerotriangulation, which now offers some new aspects with digital images (such as the automatic extraction of link points, for example). We will start with a summary of the image processing techniques seen from the point of view of the photogrammetrists, i.e. when the geometry is critical (§2.1). In the same way, the techniques of data compression are analysed for the specific case of digital images whose geometry, here too, is critical (§2.2). A summary about the use of GPS in airplanes (§2.3.1) is then presented: this is not specific to digital photogrammetry, but it deals with many practical problems of today's photogrammetry. Then a short presentation about the aerotriangulation process is proposed (§2.3.2). Then, under the title 'automatization of the aerotriangulation' is presented the automatic extraction of link points, and their use to compute the index maps (§2.4), which is the first step when processing a set of images. Then we close this chapter with a survey of the techniques used for digital photogrammetric workstations (§2.5). The automatic cue extraction methods, available at the time of completing this book, for planimetric items (the vegetation, the buildings, the roads, . . .) are still very uncertain and cannot be yet considered as operational, thus we have not presented them, as they are still in the research laboratories. Of course in Chapter 3, the automatic extraction of the altitude will be explained, with all the induced problems of DTM, DEM, DSM: this is the only operational process available that uses all the possibilities offered by digital images (and it is the simplest one, any other cue extraction requires much more human interpretation from the operator, and thus is also much more difficult to implement on a computer).

2.1 IMAGE IMPROVEMENTS

Alain Dupéret

2.1.1 Manipulation of histogram

The improvement of the image contrast is a process that tends to make the images more beautiful to use, on sometimes subjective criteria. Even if we cannot necessarily say how a good image should look, it is often possible to say if it is optimized, sufficiently, for example, to see some details. There are therefore several techniques of image improvement, but manipulations of a histogram probably represents one of the most used methods.

2.1.1.1 Definition of the histogram of a image

In statistics, the histogram is a bidimensional representation of the h function, called density of frequency, defined by:

$$h(x) = h_i \quad \text{if } x \in [a_{i-1}, a_i]$$

$$h(x) = 0 \quad \text{if } x < a_0 \text{ or if } x > a_k \quad \text{verifying} \int_{-\infty}^{+\infty} h(x) \, dx = 1. \quad (2.1)$$

In image processing, the histogram is therefore the diagram representing the statistical distribution (h_i, n_i) where h_i represents the ith level of possible colour level among the N possible values and n_i the number of pixels presenting value h_i in the digital image. By misuse of language, the h function is often called a histogram. This function is always normalized in order to be used like a law expressing the probability that a colour level n_i has to be present in the whole image. The grey level is considered therefore like a random variable that takes its values in a set of size N, being often 256 by reason of the usual coding of the elementary information of every channel on a byte. The new digital sensors, whose intrinsic dynamics ranges over several thousand levels require more space for storage.

A priori, no hypothesis can be proposed on the shape of a histogram and it is quite normal to have a multimodal histogram, presenting several attempts. A corresponding histogram to a Gaussian distribution is possible but more current in the case of satellite images. The only certainty concerns the extent, of which it is recommended that it is as large as possible, up to possibly using the 256 available levels. (See Figure 2.1.1.)

2.1.1.2 Visual representation of data

In practice, when data present a weak dynamic at the level of making their exploitation difficult, transformations intended to improve the contrast are

Figure 2.1.1 Excerpt of an aerial image and the histogram associated with this image.

still possible and will be presented. These will act by modifying coded values of the image, either while adapting the transcodage table (Look-Up-Table) used by the screen for the colour display. The user should keep in the mind also, that when an image is displayed on a computer screen, its contrast also depends on the adjustment of the monitor but also and especially on the spectral sensitivity of the eye, that presents a maximum for green-yellow colour levels; for a given wavelength variation of dλ, it implies a weaker colour level perception for the red or the blue-green, situated on the edges of the curve of spectral sensitivity of the eye, which is variable from one observer to another. Other factors are to be taken into account also, such as the prolonged observation of an intense colour provoking a transient colour aberration due to the retinal pigments, the global environment of the point observed that can make appear a more or less clear colour, as well as side effects.

Image improvements 81

2.1.1.3 Generalities on the modification of histogram

Manipulations of histograms are punctual operations that transform an input image $I(x, y)$ into an output image $I'(x, y)$ according to a law $I'(x, y) = f[I(x, y)]$. The f function is presented as a graph for which abscissas are colour levels of I, and the ordinates colour levels of I'. They are often destined to fit the dynamics of input images that are initially not contrasted enough, too clear or too dark. This redistribution of colour levels of the image can lead to some interpretation errors when the f function is non-monotonous or discontinuous:

- the horizontal sections in the graph of f lead to information losses;
- some vertical jumps provoke discontinuities resulting in false contours or aberrant isolated points;
- some negative slopes lead to local inversions of contrast.

The histogram of the image I' is centered in $ac + b$ and its width multiplied by a. In image processing software, the parameter 'brightness' can be associated to the b value and the parameter 'contrast' to the value a, slope of the curve transforming Di into Di'. (See Figure 2.1.2.)

2.1.1.4 Dispersion of the dynamics

As the raw image, considered here with pixels coded on one byte, never takes all possible values between 0 and 255, it results in a contrast default even though one knows that the eye can only discern a few tens of grey levels on a screen. Thus it is convenient to spread out the initial dynamics $[m, M]$ in the largest possible interval $[0, 255]$ with the help of the operation $D_{I'} = (D_I - m) \times [255/(M - m)]$. This operation does not provide

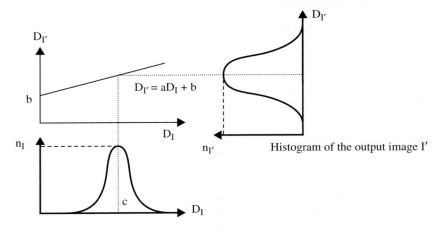

Figure 2.1.2 Manipulations of histogram.

any new information and is more a strategy of display on the screen than an interpolation technique. It is often proposed at a more elaborate level, giving better results, by calculating two levels (min and max), as

$$\int_0^{min} h(x)\,dx = \int_{max}^{255} h(x)\,dx = \varepsilon, \quad \varepsilon = 0.01 \text{ or } 0.00,$$

(h represents the function of distribution of colour level levels in the image). Then the dispersed display ranges from [min, max] to [0, 255].

An interesting variant exists when the input histogram is unimodal and relatively symmetrical and determines min and max according to the statistical properties of the colour-level distribution, as illustrated in Figure 2.1.3.

For a multispectral image, the dispersion is achieved independently on the different channels, at the risk of provoking a modification of the colour balance, each channel having initially a dynamics that is its own for a given raw image.

It is quite possible to proceed to a non-linear redispersion of the dynamics whose advantage is to reinforce contrasts for a part of colour levels of a histogram: logarithmic shape to reinforce the dark zones, exponential one for the clear zones. Often used, this strategy has the drawback of decreasing contrasts in the zones of the histogram not affected by the strong curvature of the transfer curve but it does allow a more faithful replication of the way the eye responds to different levels of brightness. (See Figure 2.1.4.) A strong slope on the central curve means an accentuation of contrasts in the corresponding zones, which are here the numerous dark parts of the right image. The thinning of contrasts in the clear parts is not here visually bothersome.

2.1.1.5 Equalization of a histogram

The equalization of a histogram is a method that looks for an equiprobable statistical distribution of levels h_i, which means to put $p_s(s) = c$ where c is a constant.

The relation $p_r(r)\,dr = p_s(s)\,ds$ being valid for all values, the search for T in $R = T(S)$ leads therefore to

$$\int_0^s c\,ds = \int_0^r p_r(r)\,dr \Rightarrow c = \frac{1}{s}\int_0^s p_r(x)\,dx,$$

which provides the sought-after relation. (See Figure 2.1.5.)

The previous formulation presumes that r and s can take continuous values. With the digital images, it may be only a finite number of values in the interval $[r, r+dr]$. When redistributing this interval on a larger domain, the principle of conservation of the number of pixels means therefore that

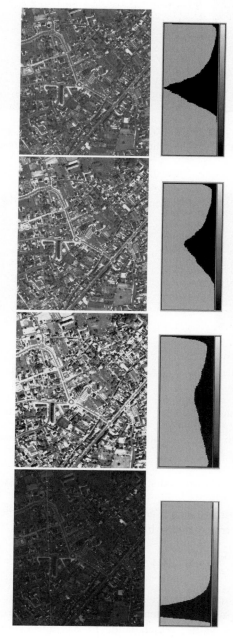

Figure 2.1.3 For a given raw image on the left, the visual results are given for an interval [min, max] being worth successively [$moy-\sigma$, $moy+\sigma$], [$moy-2\sigma$, $moy+2\sigma$], [$moy-3\sigma$, $moy+3\sigma$] toward the right.

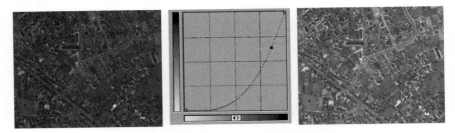

Figure 2.1.4 Non-linear redispersion of the dynamics: on the left, the original image, on the right, the modified one through the curve in the centre.

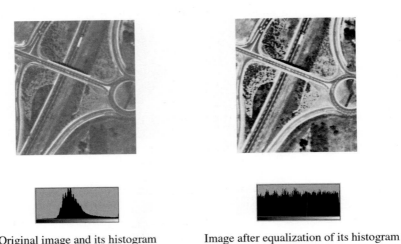

Original image and its histogram   Image after equalization of its histogram

Figure 2.1.5 Application of the method of equalization of the histogram.

only the concerned pixels are redistributed and that the histogram cannot be virtually flat.

For an image I constituted of n colour level levels quantified on N levels $D_I(k)$, these last are transformed by the law

$$D_{I'}(k) = (N - 1) \sum p_I\bigl(D_I(k)\bigr) \text{ where } p_I = \frac{n_I}{n},$$

n_I representing the number of pixels of values $D_I(k)$. As the results $D_{I'}(k)$ may be decimal, the method implemented in the software rounds off then to the nearest integer value, which leads to having several $D_I(k)$ levels associated to $D_{I'}(k)$, to which are affected the sum of probabilities of the

$D_I(k)$ from which it originates, thus justifying the non-uniformity of the histogram. To palliate this effect, some methods redistribute from pixels the more represented in colour levels toward the insufficiently represented neighbouring values, until reaching a balanced population for every quantification level.

If the number of colour levels in I' is therefore lower than in I, all possible levels not being necessarily represented, this technique allows the best possible dynamics and gives strong contrasts. Nevertheless, it may not give visually good results, colour levels of objects photographed not necessarily being suitable for such a hypothesis.

2.1.1.6 Obtaining a histogram of any shape

The operator can very well choose to have the function $p_s(s)$ as the final histogram, which means then

$$\int_0^s p_s(y)\,dy = \int_0^r p_r(x)\,dx\,,$$

which permits one to define s according to r.

Generally speaking, the following steps are achieved:

- equalization of the original image $y_1 = g(r) = \int_0^r p_r(u)\,du$;

- equalization of the desired histogram $y_2 = f(s) = \int_0^s p_s(u)\,dv$;

- application of f^{-1} (function that is calculated easily by tabulation) to the histogram $g(r)$, or $f(s) = g(r) \Rightarrow s = f^{-1}g(r)$ (the two histograms being identical).

2.1.1.7 Improvement of local contrast

The previous methods are global manipulations of the image and do not take into account the possible disparities of contrast or quality that may exist within any given image. To avoid this inconvenience, techniques of local improvement of contrast have been developed. In the neighbourhood of each pixel of size $n \times m$ that is going to be displaced progressively on all pixels of the image, the local histogram is calculated in order to be able to apply methods previously mentioned, such as the equalization or the assignment of a shape of a given histogram (e.g. Gaussian shape). As, each time, only a new column or a new line appears with the transfer of a pixel in the image, it is possible not to repeat the complete calculation of the histogram while using calculations of the previous iteration.

Thus, for a given image, the definitive assignment of a colour level for a pixel requires each time a process of manipulation of the present value histogram in a neighbourhood of given size.

The statistical process operates on sub-images of variable size and is going to use the mean intensity and the variance that are two applicable properties to model the image appearance. For a mean radiometric value $moy(i,j)$ inside a mean window centred on the pixel of coordinates (i,j), and a radiometric variance of $\sigma(i,j)$ inside the same neighbourhood, the value of the radiometry in I may be modified in I'

$$I'(i,j) = A \times [I(i,j) - moy(i,j)] + moy(i,j) \text{ with}$$

$$A = \frac{kM}{\sigma(i,j)}, \text{ scalar value called gain factor} \qquad (2.2)$$

M: mean value of the radiometries in the whole image $k \in \,]0,1[$

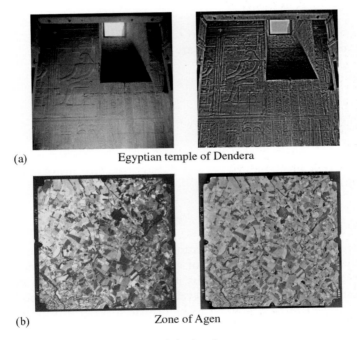

(a) Egyptian temple of Dendera

(b) Zone of Agen

Figure 2.1.6 Improvement of the local contrast.

(a) In terrestrial photogrammetry, conditions of lighting are sometimes far from optimum. Effects of over- or under-exposure are decreased by a local improvement of contrast, but beyond a given threshold, the noise of image becomes unacceptable.
(b) The image on the left presents clear zones in the top-left corner (hot spot) that can be suppressed while imposing the image in exit on the right to present a Gaussian histogram inside a window whose size is chosen by an operator.

Image improvements 87

so as to heighten the contrast locally for the weakly contrasted zones, A being inversely proportional to $\sigma(i, j)$.

The size of the processing window is a predominant criterion:

- if small (3×3 to 9×9), it permits the strongest possible improvement of the local contrasts, the side effect being to increase the image noise (see the pair of images on the Egyptian temple of Dendera shown in Figure 2.1.6);
- if large (50×50, 100×100 ...), it allows the averaging of the values of colour levels on large zones, thus decreasing the disparities of brightness in the different places of the image, but its local effects are then less visible (see the pair of images on the zone of Agen).

These methods are often used to improve the similarity of two images acquired at different dates; in aerial imagery, it allows the reduction of

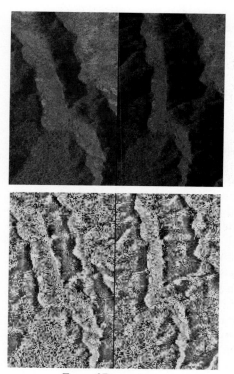

Zone of Roquesteron

Figure 2.1.7 Improvement of contrast.

Stereogram of the raw images in the upper part, and of images with local improvement of contrast on small size windows; even though the increased noise appears unacceptable on an image, the stereoscopic examination in zones of shade is nevertheless possible and limits the need to order a new aerial acquisition or the exploitation of other images.

disparities of radiometry between two distinct images when they must be merged into a mosaic, or even an ortho-image.

Beyond the simple process of a unique image, the use of a couple of stereoscopic images can be very helpful at medium or large scales, especially when zones of shades embarrass the perception of objects (see Figure 2.1.7).

The application to multi-spectral images (several channels) is more delicate, as previously seen, processes having to be applied independently to each channel to provide a coherent result. In the case of colour images, the representation can be performed in a colorimetric reference with red, green and blue main components; every coordinate of a pixel along one of these axes is the colour level it has in the corresponding spectral band. The set of points corresponding to pixels in this reference forms a cloud in which one searches the main axes along which data are best distributed. This is performed using the method of classification in main components often used in remote sensing.

The operation of local improvement of contrast can be achieved while passing in the frequency domain and using techniques of homomorphic filtering.

2.1.2 Filtering – suppression of the noise

2.1.2.1 The noise in the images

The signal in the image is composed of information, superimposed with noise; in the case of an image, it may be the noise inherent to the digital sensor used at the time of the acquisition, of the nature of the photographic document if it is a document that was digitized. Methods already presented processed every pixel independently of the neighbouring pixel value, and also have the effect of increasing without distinction noise and information in the images, which justifies the processes devoted to decreasing the noise. This noise is formed of pixels or small isolated pixel groups presenting strong variations with their environment.

In practice, images have a certain spatial redundancy, the neighbouring pixels often having some colour levels of very close values. If some general hypotheses on the noise are made (additive, independence of the signal and greatly uncorrelated), it is possible to achieve operations of filtering intended to limit the presence of the noise in images while achieving low-pass filtering, linear or not, in the spatial or frequency domains. The justification of the low-pass character of the filter is as follows.

Usually, the noise in images is not correlated with the signal level, and presents therefore a fairly uniform spectral distribution. As for images the part of low frequencies is more important than that of high frequencies, a moment happens where the signal is dominated by the noise, from where comes the idea of threshold beyond a limiting frequency. The elimination

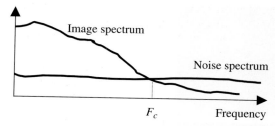

Figure 2.1.8 Beyond F_c, the spectrum of the image is dominated by the noise.

of this part of the spectra of course destroys a certain part of the signal which results in a image with little texture and fuzzier contours.

This process can by calculation of Fourier transform of the image on which is applied a process entrenching all the frequencies beyond the threshold value, then by an inverse Fourier transform. (See Figure 2.1.8.)

There are two general types of noise (see Figure 2.1.9):

- additive noise:
  - of pulse type, whose detail is to tamper randomly with some pixel colour levels to produce a 'pepper and salt' effect in images;
  - of Gaussian type, if a variation of colour level following a centred Gaussian probability density is added on colour levels of the image – this is the most current type;
- multiplicative noise, as for example that resulting from a variation of illumination in the images.

2.1.2.2 Use of filtering by convolution

Methods of filtering are numerous and are often classified by type: linear filtering, non-linear, morphological (opening, closing), by equation of diffusion (isotropic or not), adaptive by coefficients, or adaptive windows.

The use of convolution operators in the spatial domain (the image) transforms a $I(x, y)$ image into $I'(x, y)$ using a square window of odd dimension $2n+1$, moved on the image, by a matrix A with coefficients $a_{i,j}$ so that

$$I'(x, y) = \sum_{i,j} a_{i,j} I(x+i, y+j) \text{ where } -n \leq i, j \leq n. \qquad (2.3)$$

2.1.2.3 Linear filtering in the spatial domain

The simplest filtering consists in finding the average of pixels situated in a window of given size and to apply it to the central pixel, and to repeat this operation in every point of the image.

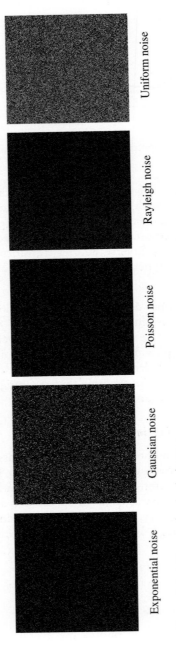

Figure 2.1.9 Types of noise in images.

The application of a filtering is a question of convolution of an image by a kernel, assimilated to the mask of the transfer function of the filter, whose coefficients can be adapted by the user according to the objective.

Examples of kernels for low-pass filters:

$$\frac{1}{9}\begin{pmatrix} 1 & 1 & 1 \\ 1 & 1 & 1 \\ 1 & 1 & 1 \end{pmatrix} \text{ or filter by the average,}$$

$$\frac{1}{16}\begin{pmatrix} 1 & 2 & 1 \\ 2 & 4 & 2 \\ 1 & 2 & 1 \end{pmatrix} \text{ or Gaussian filter.} \qquad (2.4)$$

This type of filtering is adapted to the elimination of Gaussian noise more than that of pulse noise. The way to use the filter by the average can be optimized while keeping in mind the processes applied to points of the neighbourhood; for Gaussian filter, it may be decomposed into two unidimensional filters. These artifices permit the number of operations performed in order to achieve filtering to be reduced. (See Figure 2.1.10.)

The use of such a filter decreases high spatial frequencies, as strong as the size of the window is large. One thus should be careful not to tamper with objects that one wishes to preserve in the image, because their contour becomes less and less meaningful with the progression of this filtering. Moreover, these processes generate side effects the larger the size of the window or the number of applied iterations is more important. This filtering therefore considerably degrades the contours and makes the image fuzzy, with some exceptions, as for example when the signal is stationary and the noise Gaussian.

2.1.2.4 Non-linear filtering in the spatial domain

Linear filtering has the main drawback of processing in the same way the parts of signal carrying information and noise, which sometimes justifies the use of non-linear processes, more able to attenuate the aberrant values of pixels whose colour level is too distant from the neighbouring ones. Also called filtering of rank, these methods replace the central or current pixel by one of the values selected from pixels of the V_{xy} neighbourhood of it (x, y): pixel of minimal, maximal or median value . . .

This type of method is more robust with regard to the noise of the image and has the advantage of assigning an existing value without calculating a new one.

Figure 2.1.10 Gaussian filtering.

On the left, an image digitized at 21 μm on an aerial photograph. The homogeneous zones, including noise, contain a random texture that does not constitute a useful signal. In the centre, the original image after Gaussian filter. If the noise was attenuated, contours also became fuzzy. On the right, an image of a digital camera from IGN-F: zones are naturally weakly noisy because of the nature of the sensor and its very good dynamics.

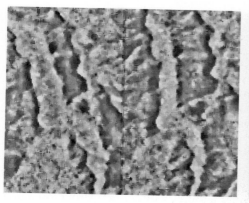 

Figure 2.1.11 Median filter.

Use of the median filter on the previously used raw images. It presents as a main defect to round corners and to smooth lines whose width is less than a half of that of the filter.

2.1.2.5 *Median filter*

In this process, the value assigned to the pixel is the statistical median value in the filtering window, whose number of elements must therefore be odd. The median is indeed the best evaluation of the mean value of a whole sample, less sensitive to aberrant values, and it is supposed here that they correspond to the extreme values of the histogram inside the mask of the filter. (See Figure 2.1.11.)

This process allows one to eliminate the isolated values that correspond to tips of pulse noise and preserve contours of objects more efficiently than the linear filters. Nevertheless, distortions misled by this filter on objects increase with the size of the window used. This filtering is therefore especially advisable when images present a noise of pulse type, and therefore important variations of weak extent. Some implementations use a mask in the shape of a cross centred on the current pixel, a shape which is preferred to the square shape to limit distortions in the image.

2.1.2.6 *Adaptive filtering*

Filters are said to be adaptive in the following cases:

- coefficients consider every term of the filter by the average, by a term that decreases with the similarity between the considered pixel and the central pixel of the window;
- the window is chosen either in shape or in size.

The filter of Nagao, relatively unknown, provides nevertheless very good results in the case of aerial images, in particular in urban sites. It is structured in four steps:

1. a window 5×5 around the central pixel is divided into nine windows of nine pixels each;
2. the first two models are each declined in four diagrams deduced from the one presented by successive rotations of 90°, 180° and 270°;
3. the homogeneity of every window is measured, with the help of an indicator of the radiometric variance type;
4. the central pixel is replaced by the most homogeneous average within the nine zones.

The filter obtained strongly smoothes textures without tampering too much with the contours of objects in the images. (See Figures 2.1.12 and 2.1.13.)

The filter of Nagao has the defect of smoothing the thinnest features of the image. Some algorithmic variants therefore exist to find the most applicable indicators of the homogeneity measure. The size of windows used, the weighting type used and the shape of masks of application can be adapted empirically in such a way as to get the best filter for the preservation of the lines, the corners or the suppression of the noise.

Figure 2.1.12 Filter of Nagao, first step.

Figure 2.1.13 Filter of Nagao applied (right) to the left image.

2.1.2.7 Filtering in the frequency domain

The filter in the frequency domain takes place in three steps:

- calculation of $F(u, v) = TF(I(x, y))$; in general, it is fast Fourier transform that is used on the base image;
- multiplication by $H(u, v)$, transfer function of a ad-hoc filter: $G(u, v) = H(u, v) \cdot F(u, v)$;
- obtaining the image filtered by inverse Fourier transform of $G(u, v)$.

The time of computation becomes a major parameter that must be considered, before any use in an operational context. Practically sometimes one uses, instead of direct and inverse Fourier transform, a square mask whose coefficients are the discrete elements of the transfer function of the frequency domain filter. Butterworth filters, low-pass exponential and Gaussian are the more frequently evoked, even though, for the meantime, only few applications in digital photogrammetry use them.

Homomorphic filtering is also usable to correct irregularities of lighting of an object. Note that a $I(x, y)$ image can be written as the product of $e(x, y) \cdot R(x, y)$, where e is the function of lighting that rather generates some of the low frequencies at the time of the calculation of Fourier transform, and R is the reflectance that rather generates some high frequencies by the nature of objects that composes it.

The two effects are made additives by creating a logarithmic image so that:

$$\ln(I(x, y)) = \ln(e(x, y)) + \ln(r(x, y)). \qquad (2.5)$$

Homomorphic filtering will increase the high frequencies and reduce the low frequencies to the point of reducing variations of the lighting while details will be reinforced, thus permitting a better observation in the dark zones of the image (see Figure 2.1.14). The function of transfer is the shape

$$H(\omega_x, \omega_y) = \frac{1}{1 + \exp(-s\sqrt{\omega_x^2 + \omega_y^2} - \omega_0)} + A,$$

with values as $s = 1$, $\omega_0 = 128$ and $A = 10$, these parameters being joined as follows to the Γ_H and Γ_L parameters and by

$$\Gamma_L = \frac{1}{1 + \exp(s\omega_0)} + A, \quad \Gamma_H = 1 + A.$$

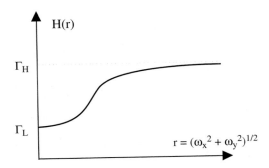

Figure 2.1.14 Homomorphic filtering.

2.1.3 Improvement of contours

Points of contours form the majority basis of the information included in the images, probably more than textures, which play nevertheless an essential role, in particular for the perception of the relief. The recognition of an object is performed most often only from its contours. The difficulty resides in the subjective notion that the user presumes to define the utility and the importance of a contour. The multiplicity of contour detectors results from the variety of the studied images and from the type of application.

2.1.3.1 Methods of the gradient

This first category is based on the use of the first derivative of the luminous intensity. The most often met operators are constructed below from models:

$$\begin{pmatrix} 1 & -1 \\ 0 & 0 \end{pmatrix} \begin{pmatrix} 1 & 0 \\ 0 & -1 \end{pmatrix} \begin{pmatrix} 1 & 0 & -1 \\ 1 & 0 & -1 \\ 1 & 0 & -1 \end{pmatrix} \begin{pmatrix} 1 & 0 & -1 \\ 2 & 0 & -2 \\ 1 & 0 & -1 \end{pmatrix} \begin{pmatrix} 5 & 5 & 5 \\ -3 & 0 & -3 \\ -3 & -3 & -3 \end{pmatrix}.$$

Gradient   Roberts   Prewitt   Sobel   Kirsh   (2.6)

The three last operators have the advantage of being less sensitive to the noise than the traditional derivative, but unfortunately also create more drifts. As each are directional, it is usual to construct similar filters to detect the contours in the eight possible directions.

The convolution by Sobel operators

$$\begin{pmatrix} -1 & 0 & 1 \\ -2 & 0 & 2 \\ -1 & 0 & 1 \end{pmatrix}, \begin{pmatrix} -2 & -1 & 0 \\ -1 & 0 & 1 \\ 0 & 1 & 2 \end{pmatrix}, \begin{pmatrix} -1 & -2 & -1 \\ 0 & 1 & 0 \\ 1 & 2 & 1 \end{pmatrix}, \begin{pmatrix} 0 & -1 & -2 \\ 1 & 0 & -1 \\ 2 & 1 & 0 \end{pmatrix} \ldots$$

(that would be to the number of eight by displacing the whole coefficients by a quarter of a turn for every new filter) allows one to detect zones of contours in a given direction.

The superposition of the four images obtained with the above-proposed filters achieves an image of the amplitude of the gradient; for every pixel $I(x, y)$, the maximum found in the intermediate images of coordinates (x, y) is selected. The presence of a contour can be decided in relation to a threshold on this value. (See Figure 2.1.15.)

It will often be interesting to reduce the noise of an image with methods like the median filter or the filter of Nagao, which will not prevent the contours from suffering from an approximate quality if they are too noisy, thick and non-continuous.

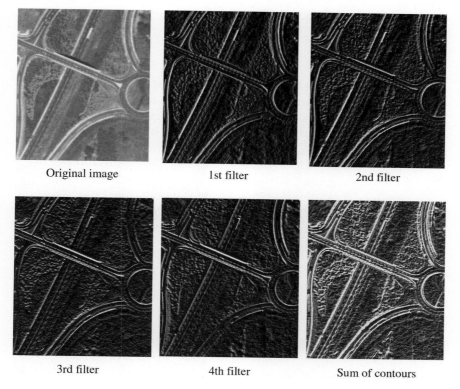

Figure 2.1.15 Sobel filter.

2.1.3.2 Methods of passage by zero of the second derivative

This second category is based on the survey of the passage by zero of second derivatives of the luminous intensity in order to detect contours. If zeroes of the second derivative constitute a closed network of lines, the second derivatives are generally very noisy. To palliate this inconvenience, it is possible to widen the support of the mask used, to select only passages by zero accompanied by a strong slope of the signal, or again to filter contours obtained after the detection.

There are several types of Laplacian:

$$\begin{pmatrix} 0 & -1 & 0 \\ -1 & 4 & -1 \\ 0 & -1 & 0 \end{pmatrix} \begin{pmatrix} -1 & -1 & -1 \\ -1 & 8 & -1 \\ -1 & -1 & -1 \end{pmatrix} \begin{pmatrix} -1 & 2 & -1 \\ -2 & 4 & -2 \\ 1 & -2 & 1 \end{pmatrix}. \qquad (2.7)$$

After the application of these masks to the image, one detects contour points by the detection of passages by zero of the image obtained. The singleness of the passage by zero provides contours whose thickness is directly one pixel. (See Figure 2.1.16.)

One shows that, if the noise is Gaussian, the most suitable operator is constituted by the Laplacian of a Gaussian defined by:

$$G(x, y) = \frac{1}{2\pi\sigma^4}\left(2 - \frac{x^2 + y^2}{\sigma^2}\right) \times \exp\left(\frac{x^2 + y^2}{2\sigma^2}\right). \qquad (2.8)$$

Another method calculates passages by zero of the directional second derivative in approximating the intensity in a window by a polynomial whose coefficients are calculated by convoluting images with masks. One can calculate second derivatives in the direction of the gradient and find

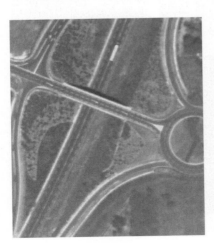 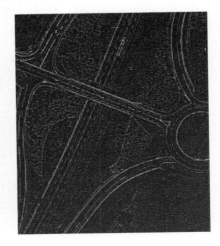

Figure 2.1.16 Use of a Laplacian filter (right) on the left image.

points of contour as passages by zero of these directional second derivatives. For a point of coordinates (i, j), the coefficients approximating the intensity in the basis

$$\left(1, i, j, (i^2-\tfrac{2}{3}), ij, (j^2-\tfrac{2}{3}), (i^2-\tfrac{2}{3})j, i(j^2-\tfrac{2}{3}), (i^2-\tfrac{2}{3}), (j^2-\tfrac{2}{3})\right)$$

are the following:

$$\frac{1}{9}\begin{pmatrix} 1 & 1 & 1 \\ 1 & 1 & 1 \\ 1 & 1 & 1 \end{pmatrix}, \frac{1}{6}\begin{pmatrix} -1 & -1 & -1 \\ 0 & 0 & 0 \\ 1 & 1 & 1 \end{pmatrix}, \frac{1}{6}\begin{pmatrix} -1 & 0 & 1 \\ -1 & 0 & 1 \\ -1 & 0 & 1 \end{pmatrix}, \frac{1}{6}\begin{pmatrix} -1 & -1 & -1 \\ -2 & -2 & -2 \\ 1 & 1 & 1 \end{pmatrix},$$

$$\frac{1}{4}\begin{pmatrix} 1 & 0 & -1 \\ 0 & 0 & 0 \\ -1 & 0 & 1 \end{pmatrix}, \frac{1}{6}\begin{pmatrix} 1 & -2 & 1 \\ 1 & -2 & 1 \\ 1 & -2 & 1 \end{pmatrix}, \frac{1}{4}\begin{pmatrix} -1 & 0 & 1 \\ 2 & 0 & -2 \\ -1 & 0 & 1 \end{pmatrix}, \frac{1}{4}\begin{pmatrix} -1 & 2 & -1 \\ 0 & 0 & 0 \\ 1 & -2 & 1 \end{pmatrix},$$

$$\frac{1}{4}\begin{pmatrix} 1 & -2 & 1 \\ -2 & 4 & -2 \\ 1 & -2 & 1 \end{pmatrix}. \tag{2.9}$$

Sometimes the computation cost of these latter processes is considered too high, so that approaches of compromise between speed and performances have been developed; thus for example the Deriche filter, which possesses an infinite pulse response of the shape $f(x) = -c \times \exp(-\alpha|x|)$ permitting therefore an implementation by means of separable recursive filters. 2D filtering is obtained by the action of two filters crossed in x and y, and the filtering of the noise with the help of a function of extension (generally a Gaussian of the same standard deviation as the one considered for the detector). (See Figure 2.1.17.)

Figure 2.1.17 From left to right: image of contours by a Laplacian, original image, and contours by the gradient of Deriche.

2.1.4 Conclusion

The image improvement can often be performed by techniques of convolution with the suitable filters, of finite extension, and thus that permit an implementation on the image as linear operators. The improvement of an image aims at improving its aesthetic through subjective criteria.

The improvement of the image may also use:

- techniques of manipulation of colour-level levels using properties of the histogram to which the user can give the shape that he wants;
- techniques of filtering in order to reduce the noise and to improve the contours in the image. The most frequently used methods are non-linear.

Bibliography

Haralick R.M. (1984) Digital step edges from zero crossing of second directional derivates. *IEEE Transactions on Pattern Analysis and Machine Intelligence*, vol. PAMI 6, no. 1, January, pp. 58–68.

Deriche R. (1987) Using Canny's criteria to derive a recursively implemented optimal edge detector. *The International Journal of Computer Vision*, vol. 1, no. 2, May, pp. 167–187.

2.2 COMPRESSION OF DIGITAL IMAGES

Gilles Moury (CNES)

2.2.1 Interest of digital image compression

The digitization of pictures offers a large number of advantages:

- possible processing by powerful software;
- reliability of the storage (on CD-ROM, hard drives ...);
- errorless transmission (thanks to the error-correcting codes).

Nevertheless, it has the inconvenience of generating often large volumes of data. To give examples, a spatial remote sensing picture acquired by the Spot satellite in panchromatic mode represents a volume of: 6,000 lines times 6,000 columns at 8 bits/pixels = 288 Mbits. A classical digitized aerial image, scanned with a 14 μm pixel size, provides 2,048 Mbits.

Of course the first goal of the engineer, when designing the data flow system, should be to define the number of effective bits by pixel through a careful consideration of the noise in the image, so that the expected noise level is at the level of the last bit, for example. But quite often the final figure is larger, in order to be able to cope with situations significantly different from the nominal one, which means very often one or two bits beyond what should be optimal.

Considering limitations that apply in most systems on capacities of storage and/or transmission, it is necessary first to reduce to the minimum the quantity of necessary bits per pixel to represent the picture. To compress, one chooses the representation that provides the minimum number of bits, while preserving in the picture the necessary information for the user. The efficiency of the compression will be measured by the rate of compression, that is the ratio between the number of bits of the picture source to the number of bits of the picture compressed.

The compression of pictures is possible for the following reasons:

- Data pictures issued from the instrument of image acquisition present a natural redundancy that doesn't contribute to information and that it is possible to eliminate before storage and transmission. One can compress pictures efficiently therefore without any loss of information (so-called reversible compression). Nevertheless, we will see in the following that this type of compression is limited to relatively low rates of compression (typically between 1.5 and 2.5 according to the content of the image).
- Besides, the end user of pictures is interested, in general, in only part of the information carried by the picture. It is what one will call the relevant information. It will therefore be possible to compress pictures more efficiently again while removing non-relevant information (so-called compression with losses) and this to the same satisfaction of the user. Rates of compression will be able to reach much higher figures (up to 50 in certain applications of very low quality). Of course, the level of distortion due to the compression will be a function of the rate and the performance of the algorithm used.

For every application, there will be a compromise to find, according to limitations of resources (storage, transmission), between the satisfaction of users (the quality of pictures being inversely proportional to the rate of compression used) and the quantity of pictures that can be stored and/or transmitted.

2.2.2 Criteria of choice of a compression algorithm

Criteria to take into account for the choice of an algorithm of compression and a rate of compression are very varied and depend in part on the targeted application. Among the most generic, one can mention:

- The type of images: pictures of photographic type including a large number of levels of gray or colour (8 to 24 bits by pixel), or artificial pictures including only a few levels of gray (e.g. binary pictures of type fax). These two types of pictures require very different algorithms. For normalized algorithms, one can cite: the JPEG standard for the photographic pictures, the JBIG standard for the artificial pictures with few gray levels and the standard T4 and/or T6 for the black and white fax. We will now consider only the pictures of photographic type.

- The level of quality required by the user: this ranges from a strictly reversible compression need (case of certain applications in medical imagery) to needs at a very low quality level (case of certain transmission applications of pictures on the Internet). The tolerable loss level and nature will have to be refined for example with the user through validation experiments for which several levels and types of degradation will be simulated on the representative pictures of the application. The type of degradation (artefacts) is going to be rather the function of the algorithm, and the level of deterioration the function of the compression rate.
- The type of algorithm: normalized or proprietary. Principal advantages of normalized algorithms are of course compatibility and permanency, with the ascending compatibility guarantee as well. The major inconvenience is the slow evolution of norms that means that a finalized norm is very rarely the the most effective solution available.
- The type of transmission or type of access to the decompressed picture: sequential or progressive. In the first case, the picture is transmitted in block to maximal resolution; in the other case, one first transmits a low-resolution version (therefore very compact) that permits the user, for example, to choose in a data base and then select a full resolution version. Algorithms using transformation (DCT, wavelets . . .) permit, among others, progressive transmissions.

2.2.3 Some theoretical elements

2.2.3.1 *Correlation of picture data*

In a natural picture (as opposed to an artificial grid), the correlation between neighbouring pixels is very high and decreases according to the Euclidian distance between pixels. The correlation decreases very quickly with the distance and often becomes negligible to 5–10 pixels of distance. To compress a picture efficiently, and therefore to avoid the need to code and to transmit several times the same information, the first operation to achieve is always to locally decorrelate the picture. We will see several types of decorrelator in the following. The simplest decorrelation to achieve consists, instead of coding every pixel $p(i,j)$ independently of its neighbours, in coding the difference: $p(i,j) - p(i,j-1)$ with i an indication of line and j an indication of column.

2.2.3.2 Notion of entropy

The entropy is a measure of the quantity of information contained in a set of data. It is defined in the following way:

- consider a coded picture on k bits/pixel, in which every pixel can take 2^k values between 0 and 2^k-1;
- the entropy of order 0 of the picture (denoted $H_0(S)$) is given by the formula:

$$H_0(S) = \sum_{i=0}^{2^k-1} P_i \log_2 \frac{1}{P_i}, \qquad (2.10)$$

where P_i is the probability that a pixel of the picture takes the i value.

The theory of information (see reference [1]) shows that $H_0(S)$ gives, for a source without memory and therefore decorrelated, the mean minimal number of bits per pixel with which it is possible to code the picture without losing information. In other words, the maximal compression rate that it will be possible to reach in reversible compression is given by: $CR_{max} = k/H_0(S)$.

To illustrate the notion of entropy, let's take four different picture types:

1. a uniform picture where all pixels have all the same value, $H_0(S) = 0$, this picture contains no information;
2. a binary black and white picture of the type of those processed by fax machines:

$$H(S) = P_{white} \log_2 \frac{1}{P_{white}} + P_{black} \log_2 \frac{1}{P_{black}}. \qquad (2.11)$$

   One has $H(S) < 1$. In practice, $P_{black} \ll P_{white}$ therefore $H(S) \ll 1$, which explains that the reversible compression algorithms used in fax machines have mean rates of compression greater than 100;
3. the picture of Spot satellite on Genoa on 8 bits in Figure 2.2.6: $H_0(S) = 6.97$;
4. a picture (8 bits) of saturated white noise in which all values are equiprobable (flat histogram), $H_0(S) = 8$. It is not therefore possible to compress this picture in a reversible way.

2.2.3.3 Reversible compression

The reversible compression is limited by the compression rate $CR_{max} = k/H_0(S)$. To achieve a reversible compression, one first decorrelates the picture in order to reduce the entropy $H_0(S)$. One can illustrate the effect of the decorrelation on the entropy, on the picture in Figure 2.2.6:

- entropy of the picture source: $H_0(S) = 6.97$;
- entropy of the picture after decorrelation of type $[p(i,j) - p(i,j-1)]$: $H_0(S) = 5.74$.

The more efficient the decorrelation, the weaker will be the entropy of the picture after decorrelation and the higher will be the rate of compression attainable by reversible coding of data decorrelated. Nevertheless, rates of compression attainable remain relatively low (of the order of 1.5 to 2.5 on pictures of remote sensing of the type from the Spot satellite).

2.2.3.4 Compression with losses

When one searches for some compression rates higher than CR_{max}, which is the most current case, one is obliged to introduce losses of information in the chain of compression. These losses of information are achieved in general by a quantification of decorrelated data. For a given picture and algorithm of compression, there is a relation between the achieved rate of compression and the errors introduced in the picture by the compression/decompression. This curve, called 'distortion/rate' has the typical shape given in Figure 2.2.1.

The distortion is often measured quantitatively by the standard deviation of the compression error, given by the formula:

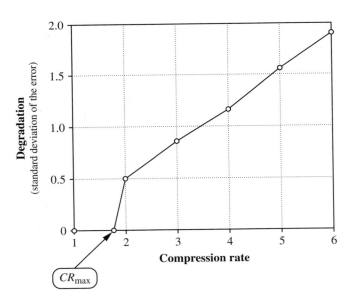

Figure 2.2.1 Typical 'distortion rate' curve.

$$\sigma = \sqrt{\frac{1}{NM} \sum_{i=1}^{N} \sum_{j=1}^{M} (p(i,j) - p'(i,j))^2} \quad (2.12)$$

for an image with N lines, M columns and a null mean error of compression, which is the most general case ($p'(i,j)$ being the value of the pixel rebuilt after compression/decompression).

The maximal level of tolerable distortion by the user of pictures fixes the maximal usable rate of compression on the mission. To compare performances of two algorithms of compression, one should draw the 'distortion/rate' curves of the two algorithms and this on the same image(s), as the performance σ (rate) of an algorithm vary considerably from one picture to another according to the entropy of data to compress.

2.2.4 Presentation of various types of algorithms

A complete presentation of the various types of compression algorithms can be found in references [1], [2] and [3]. We will give the general principles below and describe some algorithms among the more used.

2.2.4.1 General architecture

Any compression system can be analysed in three distinct modules (see Figure 2.2.2): the decorrelation of the source picture, the quantization of decorrelated values and the binary code assignment.

1. *The decorrelator* allows the redundancy of data to be reduced. In practice, there are a large number of methods, more local, to decorrelate the incoming pixels. We give some examples farther on (DPCM, DCT and wavelet transform). This phase of picture processing is perfectly reversible. At the end of this first phase, the process of compression hasn't in fact begun, but data were decorrelated so that the following processes (quantization, coding) are optimal.
2. *The quantizer* is the essential organ of the compression system. Indeed, here the quantity of information transmitted is in fact going to decrease, while eliminating all the information not relevant (in regard to the utilization that is made of pictures) included in the data coming from the decorrelator. Let's recall that non-relevant information is a quantity that only depends on the envisioned application(s). For example,

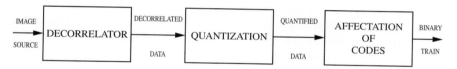

Figure 2.2.2 General diagram of a compression system.

in the case of the Spot satellite missions, radiometric precision is not relevant in strong transition zones (very contrasted contours). The quantity of non-pertinent information eliminated by the quantizer must be able to vary according to the application. We see therefore that the quantizer plays the role of organ of command of the compression system while allowing it to be used at different rates. The quantization is the only non-reversible operation of the compression chain.

3   *The assignment of codes* (or more simply coding) has the role of producing a binary stream, representative of the quantized values, which will be transmitted effectively or stored for later transmission. The rate of compression really reached by the system can be valued in a realistic way only at the end of this module. Its role is to assign to every quantized value or event a binary code that will be able to be decoded without ambiguity by the decoder. This assignment can be made more or less economic in terms of number of bits transmitted. The most efficient codes are codes of variable length whose principle is simple: to assign the shortest codes to the likeliest values (or events). An example of variable-length code is the Huffman code, used in the standard JPEG and MPEG. The implementation of the various methods of compression will not show the three already mentioned modules. Some modules will be able to be regrouped in only one (as is the case for quantization and coding in algorithms using the vectorial coding). The quantization module can also disappear as in the case of a reversible compression method.

2.2.4.2 *Reversible algorithms*

There are two main classes of algorithms to achieve the reversible compression of pictures: the universal algorithms capable of compressing any type of data, and algorithms specifically optimized for pictures, these last giving performances 20 to 30 per cent better in term of compression rate.

Universal algorithms

The most widely used is the Lempel–Ziv algorithm (LZ, LZ77, LZW) used in the utilitarian zips, gzip, and pkzip and in formats of picture tif, gif, and png. Its principle consists in marking sequences of symbols (characters of a text, values of pixel) that repeat in the file. These sequences are then stored in a dictionary that is brought dynamically up to date. Every sequence (of variable length) is coded by a code of fixed length dependent of the size of the dictionary (e.g. 12 bits for a dictionary of 4,096 elements). This type of algorithm derives benefit from the correlation of the source and adjusts itself in real time to the local statistics (by updating the dictionary). Nevertheless, the coding of pixels being linear along the scan line, one doesn't benefit from the vertical correlation in the picture, and then the performance is not as good as that of specific algorithms that use correlation in the two dimensions.

Specific algorithms

The most effective algorithm is the new JPEG-LS norm [4]. The block diagram of this algorithm is given in Figure 2.2.3. It uses a 2-dimensional decorrelator of predictive type (the most often used in lossless algorithms). The value of the current pixel x is predicted from a linear combination of pixels a, b, c (see Figure 2.2.3) previously encoded. The error of prediction is then coded with the help of an adaptive Huffman code according to the context (a, b, c, d) of the current pixel (analysed by the module of context modelling). A coding by area is used in completely uniform zones.

On the Spot satellite image of Genoa (Figure 2.2.6) one gets the comparison between universal algorithms and specific ones shown in Table 2.2.4. Thus we may achieve a gain of about 30 per cent on the rate.

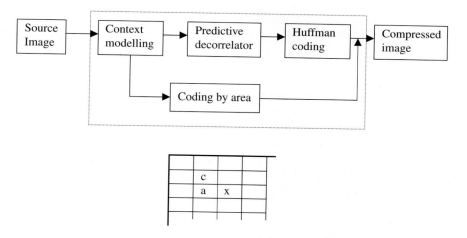

Figure 2.2.3 Block diagram of the JPEG-LS algorithm and context for the modelling and the prediction.

Table 2.2.4 Various compression rates for 2 algorithms on the Spot image of Genoa

Algorithm	Rate of compression
JPEG-LS	1.65
Lempel-Ziv (gzip-9)	1.27

2.2.4.3 *Algorithms with losses*

These algorithms are classified, from the simplest to the most complex, according to the three main types of decorrelators used: differential (DPCM), cosine transform (DCT), wavelet transform (multiresolution).

DPCM

The differential decorrelator DPCM consists in predicting the value of the pixel to code from a linear combination of values of its already coded neighbours (see Figure 2.2 3.). The prediction error is then quantized with the help of a non-uniform law adapted to the Laplace statistics of this prediction residual. This type of algorithm, used on Spot satellites 1/2/3/4 to a rate of 1.33, has the feature of being very simple and to concentrate errors of compression in strong radiometric transition zones (well contrasted contours), zones in which the absolute value of the radiometry is of low importance in remote sensing. Its disadvantage is its poor performance in terms of 'distortion/rate' for rates ranging from middle to high (>5) (see Figure 2.2.10), owing to a too local decorrelation of the signal. For the compression with losses, the DPCM decorrelator has therefore been abandoned to the profit of the DCT.

DCT

The DCT (discrete cosine transform) is the decorrelator most commonly used in compression of pictures. It is the basis of many standards of compression, among which the best known are:

- ISO/JPEG for still pictures ([5]);
- CCITT H261 for video conferencing;
- ISO/MPEG for video compression.

The DCT is one of the unitary transforms (see [1] for greater precision on the definition and the interest of unitary transforms in compression) used in compression to achieve the decorrelation. This transformation operates on blocks of $n \times n$ pixels ($n = 8$ in general) and achieves on these blocks a Fourier transform of period $2n$ of which only the real part is kept. The general diagram of compression based on a unitary transform DCT is given in Figure 2.2.5.

The direct DCT transformation of a block 8×8 is given by the following formula, where $p(i,j)$ are pixels of the source picture and the $F(u,v)$ are DCTS coefficients representative of the spectral content of the block at the different spatial frequencies:

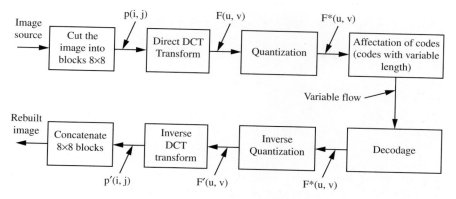

Figure 2.2.5 Chain of compression by DCT transform.

$$F(u, v) = \frac{1}{2} C_u C_v \sum_{i=0}^{7} \sum_{j=0}^{7} p(i,j) \cos \frac{(2i + 1)u\pi}{16} \cos \frac{(2j + 1)v\pi}{16}, \quad (2.13)$$

where C_u (resp. C_v) = $1/\sqrt{2}$ if $u = v = 0$ and $C_u = C_v = 1$ if not.

In the transformed domain (spatial frequency domain):

- the $F(u, v)$ coefficients obtained are correctly decorrelated contrary to source pixels of the block;
- coefficients that have a non-negligible amplitude are statistically concentrated in a restricted region of the transformed plan, which greatly facilitates the ulterior coding after quantization of these coefficients.

The quantization is a simple division by a quantification step of q. The larger the q, the higher will be the rate of compression and vice versa. The rate and therefore the error on the rebuilt data is controlled therefore by the choice of the step. At the output of the quantization, a large number of coefficients ($F^*(u, v)$) are nul. A Huffman coding is used therefore on events of type run-length adapted to this statistics. This coding, which is particularly efficient, has been optimized during the studies on the standard ISO/JPEG. This standard, of which a detailed description is given [6], is used very extensively (80 per cent of images transmitted over the Internet are compressed with JPEG).

The main drawback of algorithms with a DCT basis is due to the fact that every 8 × 8 block is coded independently from its neighbours, which creates problems of block adjusting after decompression. This phenomenon (called block effect), invisible to the eye for weak compression rates (< 8), can be bothersome for some picture-processing applications. It is illustrated in Figures 2.2.6, 2.2.7, 2.2.8 and 2.2.9 that represent respectively: the source picture, the compressed DCT/JPEG to a rate of 8 picture,

Figure 2.2.6 Excerpt of picture from Spot satellite (panchromatic mode) on the city of Genoa – non-compressed original.

Figure 2.2.7 Picture of Genoa compressed with an algorithm of JPEG type to a rate of 8 (rms = 6.8).

a zoom on the source picture and the same zoom on the compressed picture making clear the blocks effect. To suppress this block effect and to improve again the decorrelation with regard to the DCT, one has recourse to algorithms based on the decomposition in sub-bands (by wavelet transform) described hereafter.

Wavelet transform

The approach of coding techniques in sub-bands is identical to that of coding by unitary transform by block (DCT or other): analysing the signal

Compression of digital images 111

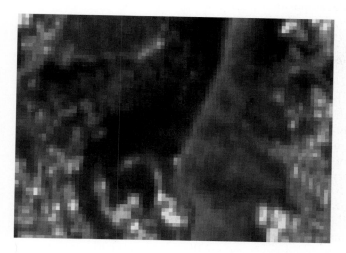

Figure 2.2.8 Zoom on the non-compressed picture of Genoa.

Figure 2.2.9 Zoom on the picture of Genoa JPEG compressed to a rate of 8 (very visible 8 × 8 block effect).

in different frequency components in order to code each of them separately. Means used to get this decomposition are, however, different: in the method by block a matrix 8 × 8 of transformation provides the different frequency components, whereas a real filtering of the whole picture must be achieved in the sub-bands technique. This decomposition in N sub-bands takes place by means of a hierarchical filtering of the picture. The filters used are determined from the theory of wavelets [7]. One thus achieves a wavelet transform of the picture.

Advantages of decomposition algorithms in sub-bands by wavelets are as follows:

- the decomposition of the picture in different spatial frequencies is made globally on the picture (by a mobile window of filtering) and not by 8 × 8 block as in the DCT. Thus there are no problems of continuity in borders of blocks on a decompressed picture;
- 'distortion/rate' performances are much better for the rates superior to 8 (typically): improvement of 30 per cent to 50 per cent of the compression rate for the same distortion level (see Figure 2.2.10);
- the progressive transmission of different resolution levels and the noise removal is very easily integrable in this type of compression scheme.

For these different reasons, the JPEG committee has been reactivated to define the JPEG2000 standard based on wavelet transform. This future ISO standard should be completed in 2000 [8] and integrates a large number of functions (random access to the compressed file, progressive transmission, resilience to transmission errors, ascending compatibility with JPEG ...).

For these algorithms based on wavelet transform, the most effective quantization/coding techniques are bit planes quantization and zero-trees coding. The state-of-the-art representative algorithms are EZWS [9] and SPIHT [10].

Comparison

One compares 'distortion/rate' performances of the three types of decorrelators (DPCM, DCT, wavelets) through three representative algorithms, on the Spot satellite picture of Genoa (Figure 2.2.6) (see Figure 2.2.10):

- DPCM: JPEG-LS algorithm in quasi-lossless mode [4];
- DCT: JPEG algorithm 'baseline' mode [5];
- wavelet: SPIHT algorithm [10].

2.2.5 Multispectral compression

The algorithms presented until now only take benefit from the intra-picture correlation (or spatial correlation). In the case of a multispectral (or colour) picture, it is also possible to exploit the correlation between the different spectral bands. In the case of the Spot satellite pictures for example, which include four bands (b1, b2, b3, mir), b1, b2 bands are very strongly correlated on certain types of landscapes whose spectral content has only moderate variations.

It is therefore interesting to achieve a spectral decorrelation then a spatial decorrelation of the picture (the inverse order is also possible). A complete

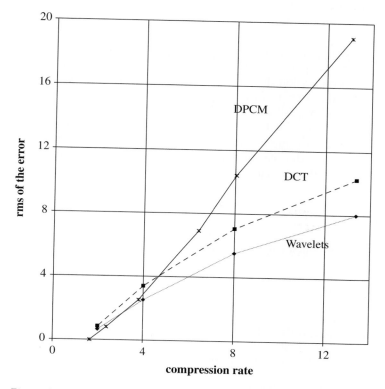

Figure 2.2.10 Comparative curves 'distortion/rate' for DPCM, DCT, wavelets.

analysis of all usable techniques can be found in [11]. The spectral decor-relator can be of predictive type or by orthogonal transform. The optimal transform to achieve this operation is the Karhunen–Loève transform (KLT). This transform enables one to switch from a spectral coordinates system [bi] where components are greatly correlated, to a reference [di] where components are decorrelated and the information (energy) of the picture is concentrated on a few components. An example of such a transformation is that used in digital television (standard CCIR 601 [12]): transfer from the system [R,V,B] to the system [Y,Cb,Cr], 80 per cent of information is concentrated on the signal of Y luminance; signals of chrominance differences (Cb,Cr) carry little information and thus can from then on be under-sampled and/or compressed to higher rates than the luminance.

2.2.6 Perspectives

Beyond the compression by wavelet transform now extensively used and which will be soon normalized (JPEG2000), different solutions are currently underway to attempt to improve the 'distortion/rate' performances of algorithms. Among these, one can cite:

- the compression by fractals [13] that, coupled to the decomposition in sub-bands by wavelet, permits progress in the high rate range (> 30);
- the selective compression that consists in detecting in the picture, zones of interest for the user, in order to apply to those a rate of compression preserving the quality of the picture, while the remainder of the picture can be compressed with a higher rate (e.g. on the Spot satellite pictures, zones of interest are zones without clouds). The selective compression opens the way to the more elevated rate used in many applications. All the difficulty lies in the reliability of the detection module.

References

[1] M. Rabbani, P.W. Jones, *Digital Picture Compression Techniques*, SPIE Press, vol. TT 7, 1991.
[2] J.P. Guillois, Techniques de compression des images, ed. Hermès, coll. Informatique, 1995.
[3] K.R. Rao, J.J. Hwang, *Techniques and Standards for Image, Video and Audio Coding*, Prentice Hall, 1996.
[4] International Standard ISO-14495-1 (JPEG-LS).
[5] International Standard ISO-10918 (JPEG).
[6] W.B. Pennebaker, J.L. Mitchell, *Still Image Data Compression Standard*, Van Nostrand Reinhold.
[7] M. Vetterli, J. Kovacevic, *Wavelets and Subband Coding*, Prentice Hall, 1995.
[8] http://www/jpeg.org (official site JPEG).
[9] J.M. Shapiro, Embedded image coding using zerotrees of wavelet coefficients (EZW), *IEEE Transactions on Signal Processing* (Special Issue on Wavelets and Signal Processing), vol. 41, December 1993.
[10] A. Said, W.A. Pearlman, A new fast and efficient image codec based on Set Partitioning in Hierarchical Trees (SPIHT), *IEEE Transactions on Circuits and Systems for Video Technology*, vol. 6, June 1996.
[11] J.A. Saghri, A.G. Tescher, J.T. Reagan, Practical transform coding of multi-spectral imagery, *IEEE Signal Processing Magazine*, vol. 12, January 1995.
[12] ITU-T Recommendation 601, Encoding parameters of digital television for studios, 1982.
[13] M.F. Barnsley, L.P. Hurd, *Fractal Image Compression*, Peters, 1993.

2.3 USE OF GPS IN PHOTOGRAMMETRY

Thierry Duquesnoy, Yves Egels, Michel Kasser

2.3.1 Use of GPS in photographic planes

Thierry Duquesnoy

2.3.1.1 Introduction

In aerial images acquisition, the GPS (global positioning system), or any other GNSS equivalent system (GLONASS, or the future GALILEO) may be used in three distinct phases:

- the real-time navigation;
- the determination of the trajectory of the plane, more precisely the determination of the position of the coordinates of the camera at the time of the image acquisition;
- the determination of the attitude of the plane.

If these three aspects use the GPS effectively, they do not require the same type of measure. The most original use concerns the measure of attitude that we will describe farther.

We will recall briefly here that two types of GPS observations exist:

- the pseudo-distances that are measures of time of propagation of the pseudo-random modulation of the signal;
- the measure of phase, performed on the wave carriers, and that is performed on an identical but stable signal generated by the receptor and locked on the satellite signal.

And two types of positioning exist, the absolute positioning and the relative positioning. The precision of the positioning varies according to the type of observations, from a positioning accurate at the level of a few metres to a relative precision of the order of 2 mm $\pm 10^{-6}$ D to 10^{-8} D using the measure of phase.

In the end of section (2.3.1.5), one will find a summary of the terminology employed for the different uses of the GPS.

2.3.1.2 The navigation

The navigation is a priori the simplest mode. It uses the pseudo-distances. At the time of the image acquisition, it is necessary to have the position, in almost real time, of perspective centres in order to verify that the follow-up of the axis by the plane is good and that the overlap rate is correct. In absolute positioning, with a precision that is about ten metres (with the SA switched off, as it is since May 2000), conditions are achieved. The metric precision realized is by far superior to the 40 m with which pilots can hold the axis of flight. If a better precision is required, the differential mode is necessary: several solutions may exist to get the differential corrections in real time. One can use networks of stations that broadcast corrections on different frequencies. One can arrange his own GPS receptor on a known fixed station that sends corrections. And then there are differential correction services via a geostationary satellite. This last process seems best suited for airborne applications.

Let's note that in differential navigation, the speed of the plane is known to 0.1 knot, which is sufficient for photogrammetric applications.

For the overlap, it may be necessary to know the attitude of the plane at the same time. We will describe these needs in the chapter dedicated to the determination of attitude.

2.3.1.3 The determination of perspective centres for the aerotriangulation

We saw that measures of pseudo-distances permit a metric positioning when differential data are available. However, the necessary precision on the determination of perspective centres must be at the same level as that necessary for the photogrammetric process, typically 10 μm × the scale of the picture. A decimetre precision is therefore necessary to most photogrammetric applications (scale varying from 1:5,000 to 1:40,000). This precision cannot be reached with the exclusive utilization of pseudo-distances. But the measure of phase is ambiguous as, while using the phase, one only measures the fractional part of the cycle, the whole number of cycles separating the receptor of the satellite being unknown.

Therefore there are two ways to improve the precision of the determination of perspective centres a priori: improve the process of the pseudo-distances while using the phase, or be able to solve ambiguities in the process of the phase.

The trajectography

This is a calculation in differential mode between a reference station of perfectly known coordinates and the antenna and receptor on board the plane. The process is based on the fact that phase data are less noisy than pseudo-distance data. Indeed, one considers that the noise of the measure is estimated to be better than 0.1 per cent of the wavelength of the signal on which the measure is made. Therefore, the noise on the pseudo-distances is metric while the noise on the phase is millimetric. Thus one tries to improve the process of pseudo-distances by decreasing the noise of measure, by a smoothing process with the help of the phase measurement. This is the trajectographic mode. In this mode no initialization is necessary. One gets by this method an internal consistency within one axis inferior to the decimetre. But then, the precision between neighbouring strips is no better than some metres.

The resolution of ambiguities on the flight

Again, this is a positioning in differential mode. In the case of two fixed receptors, the resolution of ambiguity is possible without too many difficulties for baselines of several hundred kilometres. But it is much more difficult when one of the GPS receptors is mobile.

An algorithm developed by Remondi (1991), permits the resolutions of ambiguities in flight (on the fly, or OTF) without any initialization in static phase, which would obviously not be possible.

The initialization of the ambiguities fixation in flight cannot be performed beyond a distance of 10 km between the fixed receptor and the mobile. Once the initialization is made, the basis can go until 30 km, but without cycle slip. If there is a cycle slip, an OTF traditional software will do the initialization again, and therefore the mobile receptor should come once more closer than 10 km from the reference. This method of ambiguity resolution in flight is widely used by numerous constructors and GPS users, but it has the major inconvenience, in an application of photogrammetric production, of imposing maximal lengths of bases of about 30 km. Some constructors provide solutions for bases ranging to 50 km, but no full-size test has been achieved again. Nevertheless, the order of magnitude remains the same and requires the presence of a nearby fixed receptor on the zone to photograph.

The GPS positions, calculated during the mission are those of the centre of phase of the antenna. However, the position that interests us is that of the perspective centre at the time of the image acquisition. It requires a good synchronization between the image acquisition and the GPS receiver, and a perfect knowledge of the vector antenna/camera at the time of the image acquisition. This vector is constant in the airplane frame, on the other hand it is not in the earth reference system. It is therefore necessary to calculate the matrix of passage between the two systems, either using ground points, or knowing the attitude of the plane at the time of the image acquisition.

2.3.1.4 Measures of airplane attitude

The measure of attitude by GPS is an interferometric mode. One measures times of arrival of the wave on a system of several antennas forming a strong solid shape. It requires a perfect synchronization on times of arrival to the different antennas. It is achieved with the use of receptors, a clock, and several antennas. One finds currently on the market systems of three or four antennas.

The more the antennas are separated from one another, the better is the angular precision. Unfortunately, it is very difficult to get solid polygons on planes used for aerial pictures, due to the flexibility of the wings (with the notable exception of airplanes with high wings). The currently achievable precisions are in the order of the thousandth of a radian whereas a precision ten to a hundred times superior would be necessary.

It is possible, but at a considerable cost, to couple to the GPS an inertial system (IMU, for inertial measurement unit). The utilization of the inertial system alone is not possible because it drifts very quickly. But then, coupling the inertial platform with the GPS measures will cancel most of the drifts, and makes it possible to get a precision on the angular measures of about 10^{-5} radians. This is the device used in aerial laser scanning.

Information of attitude is known in real time in the plane and can therefore also be used for the navigation in order to minimize overlap errors, and to have a sufficiently precise assembly picture at the end of the aerial mission.

2.3.1.5 GPS terminology

- *Kinematics*: differential method (DGPS) based on the measure of phase of at least four satellites. The principle consists in solving the ambiguities by an initialization, then to observe the points for a few seconds while preserving the signal on satellites during the journeys, and therefore the same integer ambiguities.
- *Pseudo-kinematics*: original method of kinematic type that consists of reobserving the reference points and treating the file of observation as if it was a fixed station having a gap in its data (ambiguities being the same).
- *Dynamic*: differential method (DGPS) with observations of pseudo-distances, the station of reference whose coordinates are known, sends by radio in real time the corrections to the mobile station, which may then calculate its position.
- *Trajectography*: specific mode that uses observations of phase and pseudo-distances, the computation being performed using pseudo-distances smoothed by the phase.
- *Navigation*: we will associate to this term the common sense of navigation, that is the knowledge of information on the position and the attitude of the plane that could allow the pilot as well as operators to rectify their trajectory or the orientation of the camera in situations requiring it.

2.3.2 Use of GPS in aerotriangulation

Yves Egels, Michel Kasser

2.3.2.1 How to limit the stereopreparation work

Whatever the photogrammetric product to be achieved (vectorial restitution, orthophotography, DTM), it is indispensable to determine in a previous phase the formulas describing the geometry of the used pictures. This phase is traditionally called aerotriangulation (or spatiotriangulation in the case of spatial images, whose geometry is often different). We will recall quickly here the principle of this classic method in analytic photogrammetry (one will find the complete survey of this problem in any good book on analytic photogrammetry), and we will develop only the improvements made possible by the appearance of digital photogrammetry.

Let's recall that the geometric properties of images alone do not permit one to know either the position, or the orientation, or the scale of the photographed objects. The determination of the corresponding mathematical similitude cannot be obtained except by external measures: thus, if one limits oneself to the case of the stereopair, and to determinations of ground control points, it will be necessary to measure at least two points in planimetry and three points in altimetry.

The aim of aerotriangulation is to reduce as much as possible this requirement of field measures, by processing simultaneously the geometry of a large number of images. The goal is to adjust (often to the least square sense) the measures of coordinates in the images of the homologous tie points, a certain number of ground points being considered as known, and possibly auxiliary measures recorded during the photographic flight as well, or satellite trajectography data. This leads generally to an overabundant system of several thousands of non-linear equations, that one will be able to solve by successive approximations from an approached solution.

Many software packages on the market allow one to perform this calculation. Without any in-flight measures, they allow the necessary ground points to be limited to one planimetric point for five to six couples, and one altimetric point by couple. The digital photogrammetry did not bring any basic difference at the level of the aerotriangulation calculation itself, but on the other hand it allows the measure of tie points, which remained manual (and very trying) in analytic plotting, to be automated almost completely. Besides, the development of digital cameras, as they require more images to cover a given zone, has led to the search for more efficient auxiliary measures in order to restrict again the ground measures.

These normal mean values (one point in planimetry for five or six couples, and one point in altimetry) depend in fact on the geometry of the work, and in particular on overlaps from image to image (the minimal value is 60 per cent), as well as overlaps between neighbouring strips that have a major impact in the rigidity brought to the general figure. If the overlap between strips is significant (e.g. 60 per cent), every point is found in three strips, but at the end it doesn't permit an appreciable decrease in the number of control points below half of values for a normal aerotriangulation. On the other hand, one gets a much better reliability, and the precision is improved as well.

This is a solution used for the survey of points that must be measured with a very high precision, materialized by targets so that the pointing is almost ideal (e.g. measure of ground deformations). On the other hand, the additional costs of this solution makes it a non-profitable one for classical cartographic studies.

2.3.2.2 Use of complementary measures during the flight

One formerly used measures of APR (airborne profiles recorder, see §1.7), a method that provided some vertical distances (radar or laser) from the plane to the ground, with as altimetric reference an arbitrary isobar surface. The airplane followed trajectories forming a grid, the points of impact of the ranging unit to the ground being controlled by pictures acquired by a camera. This methodology fell into obsolescence, but proved to be quite useful to cover vast uninhabited zones (Australian desert, northern Canada).

Use of complementary measures during photographic flights

It would ideally be necessary to know at each instant where a picture is acquired:

- the position of the centre of the perspective, that is to say the optic centre of the camera, in the reference frame of the work;
- the orientation of the camera.

These two groups of parameters are not at all equivalent in terms of necessary acquisition expense.

For the position of the centre, let us note that almost no one uses the differential barometer, which could bring some very interesting elements while creating some strong new constraints on what is one of the major weaknesses of aerotriangulation, that is to say the altimetry. This equipment is otherwise functional without any external infrastructure, unlike the GPS that we are going to examine, which is a more complete solution on the other hand.

For the orientation of the camera, there does not exist a reliable and low-priced sensor. The available inertial platforms are costly. Based on the use of accelerometers whose values are integrated twice to get the displacements, and on gyrometers or gyroscopes to measure the variations of attitude, they have by their very nature a random drift more or less proportional to the square of the time. This requires frequent updates, and this updating can then be performed by the GPS. There are also (see §2.3.1) some solutions using the GPS alone, measuring the GPS data simultaneously acquired on three or four antennas. This is a mode that works like the interferometry. Its angular sensitivity depends in fact on the antenna spacing, the values of measure of the vertical component on each antenna being hardly better than 1 cm. This implies, as the antennas cannot be more distant than 10 m (antennas on wings, the nose and the vertical stabilizer of the plane) a precision of measure that is hardly better that some millirad, with all the risks of GPS measures. By comparison, with a pointing accuracy on image around 15 μm (which is not excellent) and a classic focal distance of 150 mm, one understands that the need in absolute orientation is between 10^{-4} and 10^{-5} rad, which disqualifies de facto such a use of the GPS. Otherwise, it would be necessary that the four antennas form a rigid polyhedron, and it is not the case since most planes' wings bend according to constraints brought about by atmospheric turbulence: the only noticeable exception is formed by planes with high wings, when the implantation of antennas on wings is localized on the zone of shroud grappling that holds the wings.

2.3.2.3 Positioning by GPS

The precision that it is necessary to achieve to localize the perspective centres must be at least as good as that of points to localize on the ground, and obviously depends on parameters of the image acquisition (scale, height of flight, pixel size, etc.). In current cases it means a precision ranging from 5 to 30 cm. We are in a dynamic regime, with a speed of the plane around 100 to 200 m/s. It implies a very good synchronization between the camera and the GPS receiver, which generally does not pose any particular problem with the GPS equipment. The GPS should in this case have a special input, the external events datation being performed with a precision that reaches the 100 μs level (GPS easily performs 100 times better). But it is necessary that the camera provides a signal perfectly synchronous (better than 0.1 ms) with the opening of the shutter, which is the case with all modern cameras but is not always true of old equipment.

In addition it is necessary to get a continual positioning during the movement. With GPS measures at each second, one has to interpolate the position of the camera at the time of the signal of synchronization on distances of 100 to 200 m, which does not permit a decimetric precision as soon as the atmosphere is a little turbulent and creates perturbations on the line of flight. And one should note that the higher the requested precision, the more one flies at low altitudes and therefore the stronger the turbulence. Thus a positioning adapted to large scales requires a better temporal sampling than 1 s, a value of 0.1 s (yet rarely available on the GPS receivers) being barely sufficient. In this optics, one can associate a GPS receiver and an inertial platform that serve as a precise interpolator, the bias of the inertial unit being small on one second, so that the unit can provide data quite often (typically from 100 to 1000 Hz). But this solution is quite expensive and still not very often used.

As we saw in §2.3.1, the precision requirements need the use of the GPS in differential mode. Two modes can be used:

- *Kinematic mode*, with measure of the phase on the signal carrier. Either one initializes with a time duration permitting the ambiguity resolution (it requires some minutes) and fixes them then for the remainder of the flight. But then the reliability is low, and this is not used because when the plane turns it frequently occurs that there are short interruptions of the signal, ruptures in the continuity of the measure that would oblige us to solve the ambiguities once more, which becomes obviously impossible in flight. Either one solves ambiguities on the flight (OTF, AROF, ... methods), which overcomes this difficulty. Nevertheless, it is necessary to note that up to now these methods work only with reasonably short bases (20 or 30 km to the maximum), which is poorly compatible with work conditions of photographic planes. Indeed it would be necessary, before the beginning of the flight, to put in operation a GPS station close to the zone to photograph. Risks of photographic flight programming in the mid-latitude countries makes such a step quite restrictive, the weather alone being a sufficiently strong constraint not to add an obligation to set up such an equipment, which would lead to too much time wasted. The only cases where one can work according to this mode will be those where a permanent GPS station close to the zone to photograph exist. This method is therefore not very often used.
- *Trajectography mode*. One then measures the pseudo-distances as well as the carrier phase. The precision is not so good, but one can work with much more important distances to the reference station. We are going to study this solution in more detail.

2.3.2.4 Survey of the trajectography mode

The satellite-receptor distance can be calculated on the one hand from the time of flight of the wave deduced from the datation of the C/A code, the clock correction, the tropospheric and ionospheric corrections (typical precision of the order of one metre in differential mode). It can be calculated on the other hand from the phase measurement, with an integer k number of phase turns, from the correction of a clock, and from tropospheric and ionospheric corrections (the precision is then millimetric, except that k is unknown). The difference between these two variants of the same distance being necessarily null in theory, while integrating measures on a given satellite during enough time one can get an evaluation of k. This mode is not adapted to real-time operations, on the other hand as long as there are satellites in common visibility between the reference station and the airplane, the result is usable as far as the GDOP (geometric dilution of precision, a parameter giving an evaluation of the geometric quality of the observable satellites, the smaller figures meaning the better geometry) is correct, and the observed bases can range beyond a thousand kilometres.

Results are in general always of the same type: for a few minutes the noise of determination of airplane coordinates is lower than 5 cm, and one observes a bias, positioning errors varying slowly, with a typical time constant of 1 or 2 hours, the amplitude of the errors being of the order of one metre. This is due to the slow changes of satellite configuration in the sky, the orbit errors being directly input in the calculations.

We may then model the GPS measures as an error of general translation of the GPS positions for every applicable short temporal period, typically of 10 min, for example for every set of photographs taken along a given flight axis. One adds thus a supplementary unknown triplet in the aerotriangulation for every strip. If one works in differential mode, the precision is then centimetric, and in absolute mode this is without significance since there are reference points on the ground that completely impose the wanted translation for the whole set of images.

One thus succeeds in decreasing considerably the number of ground points: in planimetry, it is necessary to get now at least 4 points by block, and some more but only for controls, and in altimetry 1 point all 20 or 30 couples is satisfactory. These points don't necessarily need to be to the extreme borders of the zone.

One should be very attentive to problems that are linked to reference frames. In particular in altimetry one should remember that the GPS measures are merely geometric, whereas the levelling of reference is always based on geopotential measures. Geoid slopes, or rather slopes of the null altitude surface, must be taken into account according to the requested precision.

One can consider treating the work without any ground references at all (inaccessible zones for example). In this case one gets a coherent survey but with a translation error of metric size, translation that is the mean value of the systematism unknowns.

One should be also attentive to the geometry of the set of stereopreparation points. If they are all reasonably aligned, the transverse tilt will be undetermined. It is necessary to include points therefore in the lateral overlap zones between strips. If one works on a zone without sufficient equipment in levelling, one will be able to improve the quality of the aerotriangulation by crossing the parallel strips by transverse photo strips. But it will be necessary to perform levelling traverses on the ground if the geoid slope is unknown, just to impose its slope in the geometric model.

2.3.2.5 Adjusting the link between the GPS antenna to the optic centre of the camera

One will start with measuring with classical topometric methods the link vector E ranging from the antenna to the optic centre, in the airplane reference system. X_A being the position of the antenna, X_S that of the optic centre, one will write:

$$X_S = X_A + T + R \cdot E, \tag{2.14}$$

where T is the vector describing the residual systematism of these measures, and R describes the rotation of the airplane in the reference system of the study, assumed equal to the rotation of the camera, itself even deduced from the rotation of the bundle, a by-product of the collinearity equation.

2.3.2.6 Conclusion

The use of the GPS (or any other GNSS, like the Russian GlONASS or the future European GALILEO) in the plane is a very important auxiliary source of data to reduce the costs of the control points on the ground for the aerotriangulation. But in such a case it is necessary to be aware of the imperfections of the GPS, and to know what to do when some subsets of data are not exploitable. On the other hand, it must be well understood that if some inertial measurements may help considerably to interpolate within the GPS data, they cannot correct its possible defects, and thus such data should be considered as additional, every possible effort should be made to have the best possible signal in the antenna: any device allowing the absolute orientation of the camera to be provided will be welcome, but while waiting for its availability, the GPS already gives excellent results . . .

Reference

Remondi B.W. (1991) Kinematicc GPS results without static initialization. NOAA Technical Memorandum NO S NGS-55.

2.4 AUTOMATIZATION OF AEROTRIANGULATION

Franck Jung, Frank Fuchs, Didier Boldo

2.4.1 Introduction

2.4.1.1 Presentation

This section concerns the automatic determination of aerotriangulation. This automation aims at the production of tie points and their use in two domains: the aerotriangulation itself and the automatic realization of index maps, which is a preliminary step to any aerotriangulation process. The reader will note that we do not consider here the measure of the reference points. One will not treat their determination (which is made by techniques of geodesy), or the measure of their position in images (which are measured by hand).

The automatic calculation of tie points is a topic of growing interest notably by reason of the increasing number of images produced for photogrammetric studies. In particular, both the use of digital cameras that is now increasing, and of strong overlaps (notably interstrips) on the other hand, generate images in greater number than in the past. It is henceforth possible to find a substantial number of manufacturers of aerotriangulation software proposing a module of automatic determination of tie points. These modules are able to give good results on simple scenes (including neither important relief nor too large-textured zones). However, the determination of tie points in not so simple configurations still mobilizes greatly the community of researchers. These less simple configurations are notably terrestrial photogrammetric studies, aerial images of different dates, scenes with strong relief, etc.

This type of problem is a topic of a working group on behalf of the OEEPE (Organisation Européenne d'Études Photogrammetriques Expérimentales) (Heipke, 1999).

2.4.1.2 Basic notions

We should recall some notions that will be used in what follows:

- *Tie point*: 3D coordinates often corresponding to the position of a physical detail of the scene, and seen in at least two images.
- *Measure*: 2D coordinates of the projection of a point (of reference, or of link) in an image.
- *Point of interest*: a point of the image around which the signal has specific characteristics, such as high values of the derivatives in several directions or at least two orthogonal directions, and detected by a particular tool (e.g. detection of corners, of junction).
- *Similarity measure*: function associating to two neighbourhoods of two points a real finite number. The more often used measure is the linear correlation coefficient.
- *Homologous points*: set of points satisfying some properties in regard to the similarity measure. Generally, a point P_1 of an image I_1 and a point P_2 of an image I_2 are judged homologous if, for any P of I, P_1 and P_2 are more alike than P and P_2, and for any Q of I_2, P_1 and P_2 are more alike than P_1 and Q.
- *Repeatability*: quality of a detector of points of interest capable, for a given scene, of detecting points of interest corresponding to the same details in the different images, even if conditions of image acquisition vary (lighting, point of view, scale, rotation ...). A detector of points of interest is more repeatable if it produces the same sets of points for a given scene in spite of variations of the condition of image acquisition. This notion is specified in Schmid (1996).
- *Disparity*: for two images possessing a weak rotation angle one in relation to the other, the disparity of two points is the vector equal to the difference of their coordinates. In the case of aerial image

acquisitions with vertical axis, the disparity of two points corresponding to the same physical detail of the land is bound directly to the altitude of this detail. This notion may be extended to the case of images possessing any relative rotation.

2.4.2 Objectives of the selection of points of interest

This section discusses three points concerning objectives of methods of automatic measurement of tie points. We will treat first their reliability, then their precision, as well as a third important point, which makes an important difference between the automatic measurement of tie points and the manual techniques: it is about the number of tie points.

2.4.2.1 Reliability

Most methods of aerotriangulation depend on an optimization using a least square adjustment technique. This technique is especially sensitive to aberrant values. As usual it is therefore necessary, during the calculation of tie points, to try not to provide any aberrant measures.

The first major objective toward a method of automatic measurement of tie points is thus to provide points that are exempt from mistakes.

2.4.2.2 Precision

This second major objective can be the carrying out of complementary studies as soon as one knows that the aerotriangulation work has been performed correctly (i.e. there are no more mistakes).

To get a better precision for the aerotriangulation implies more precise measures of tie points, but also taking into account in methods of aerotriangulation an error model adapted to the real errors made by measuring tools (which implies a study of these errors to model them correctly). Indeed, error models used are generally Gaussian, and it remains to prove that this model is suitable for a particular measuring tool. If necessary an adaptation of the error model may be advisable.

2.4.2.3 Number of tie points

Traditional aerotriangulation uses 15 measures by image, whereas tools of automatic measure of tie points are able to provide a large amount of data.

In the manual case, these points are very satisfying, because they are selected by an operator. The operator can notably a priori assure their geometric distribution. In the automatic case, the lack of intelligence of any computer method may be partly compensated by the abundance of data. Concerning the distribution of points, it is difficult to force the machine to 'find' a solution in a given zone, and it is therefore probably preferable to let it find another solution in the neighbourhood of the desired area. Concerning errors, the abundance of data does not reduce the error rates of the methods, but permits the use of statistical tools aiming to eliminate these errors.

Studies are not however advanced enough to permit one to fix an ideal value of measures by image. Nevertheless, one expects that the number of required points be appreciably higher in the automatic case (typically 100 measures by image (Heipke, 1998)). The debate on this question remains very open.

2.4.3 Methods

The automatic measurement of tie points generally requires two major steps. The first concerns the detection and the localization of points of interest in images. Points of interest can be calculated individually from the images. Then these points are used to produce the actual tie points. The passage of points of interest to tie points can be generally performed with two opposite strategies.

In the two cases, strategies search between images for points whose value of similarity measure used is maximal. Strategies differ according to the zone in which one looks for the maximum.

For a point of interest P of an image I, the first strategy consists in searching in an image J for the position of the point Q that maximizes the value of similarity among all possible positions (all pixels of a region where one expects to find Q). One can even consider calculating the position of Q with a sub-pixel precision. This strategy, which one will call 'radiometric' is oriented: I does not play the same role as J in this case.

The second strategy consists in restricting the research space in J to only the points of interest calculated. One will call this approach 'geometric' because it is focused on the position of tie points.

2.4.3.1 Methods of detection of points

Detectors of points use particular features of the signal. We will more especially mention the detector of Förstner (Förstner and Gülch, 1987) (much used in commercial software) as well as the detector of Harris whose repeatability is high (Harris and Stephens, 1988; Schmid, 1996).

These detectors generally have a poor localization precision. Experiences have shown a shift (rms) of 1 to 1.5 pixel between the real corners and the detected points (Flandin, 1999). Nevertheless, the delocalization of these points possesses a very strong systematism but a low standard deviation (around ½ pixel). Thus it is possible to consider these points as satisfactory for a photogrammetric set-up. A better localization of these detected points can be considered by the use of specific techniques (see §2.4.2.2).

Figure 2.4.1 Two 300×200 images and their points of interest.

Figure 2.4.1 represents two neighbouring images (size 300×200) with, in white, their points of interest. Considering the density of points, their representation is difficult and the visual inspection of the images must be done carefully. Figure 2.4.2 shows matching points between these two images. An attentive observation shows that the matching is quite often correct. Obviously homologous points are in the overlap zones of the two images. Figure 2.4.3 presents two excerpts of these images, allowing one to see the details of the matching points in a small area of the images.

Figure 2.4.2 Results of the matching.

2.4.3.2 Calculations of homologous points

- Photometric aspect: a similarity criterion much used in many matching techniques is the coefficient of correlation. This coefficient is calculated on two imagettes. The advantage of this technique resides in its ability to match two zones with similar grey levels. Nevertheless, it is necessary to use perfectly oriented images (problem of images with strong rotations).
- Signal aspect: a criterion based only on the local resemblance (photometric criterion) is not always sufficient. Indeed, the correlation within a homogeneous zone or along a contour remains a source of ambiguity. A competing or complementary strategy consists in only

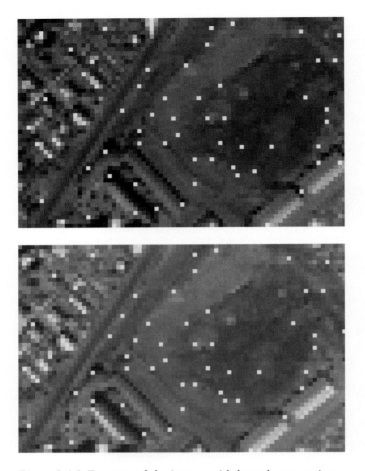

Figure 2.4.3 Excerpts of the images with homologous points.

matching the points possessing particular features of the signal (Harris and Stephens, 1988; Förstner and Gülch, 1987).

A mixed approach takes full advantage of the two methods. The detection of points is made according to features of the signal and the matching is done according to a photometric criterion.

2.4.3.3 Reliability

Tools contributing to obtaining a good reliability of tie points are numerous:

- *Repeatability of the detector*: one preferably has to use a point detector with a good repeatability. Examples are the detectors of Förstner (Förstner and Gülch, 1987) and Harris (Harris and Stephens, 1988), among which Schmid (1996) shows that Harris's possesses the best properties of repeatability.
- *Multi-scale approach*: this approach allows one to guide the research of the homologous point in full resolution with the help of a reliable prediction of the disparity on sub-sampled images. This approach allows two problems to be solved: on the one hand the problems of outliers error (one may limit the size of research zones voluntarily); on the other hand this approach allows a limitation of the combinatories of the problem of matching in full scale. One can note that the introduction of a digital terrain model (DTM) can partially replace the use of sub-sampled images to predict research zones at full scale.
- *Multiplicity*: a visual assessment of the quality of tie points permits one to note that the percentage of erroneous tie points decreases appreciably with the order of the point (number of images for which the point is visible). There are nevertheless several ways to exploit this observation. A very reliable way is to consider a multiple point as valid if all associated measures are in total interconnection with regard to the similarity function, i.e. if each pair of measures is validated by a process of image by image matching. Figure 2.4.4 shows this mechanism: every couple of points of this multiple link has been detected as pairs of homologous points by the matching algorithm.

It is necessary to note that the points with a strong multiplicity are necessary for the calculation of aerotriangulation.

Photogrammetric filtering

Forgetting any consideration of radiometric type, one can also use the photogrammetric properties of tie points. Indeed, each of these points are supposed to represent the same ground point. Therefore, all perspective rays must cross in one point. Several methods can be proposed to use this

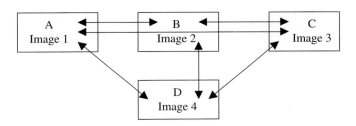

Figure 2.4.4 Example of multiple point of 4th order in total interconnection.

property. The first consists in attempting to set up every couple individually, for example by the eight points algorithm (Hartley, 1997). Technically, this algorithm calculates the fundamental matrix associated to a couple of images. If one notes u and u' the projective vectors representing, in the reference of the image, two homologous points, the fundamental matrix F is defined by: $u'^T F u = 0$. In return for some numeric precautions, the algorithm permits the calculation of the likeliest fundamental matrix associated to a sample of points. From the fundamental matrix, it is possible to determine the epipolar line associated to a point u, and therefore also to calculate the distance between u', homologous of u, and this epipolar line. This distance is a sort of 'residual' thus making it possible to qualify points. This algorithm is very general, and permits a setting up of whatever is the configuration. It has been used therefore for all sets of two images having at least nine points in common, whatever their respective positions. In this practice, for every 'couple' one attempts an iterative stake in correspondence: one sets up, one calculates residues, one eliminates points having a residual greater than three rms, and one repeats the setting up, and this until this one is satisfactory, or the process is consolidated. If the convergence could not have taken place, one then tries a stochastic approach algorithm ([Zhang and Gang, 1996]): one chooses N samples of tie points and applies the algorithm on every sample. Residues are then calculated on the whole set of points, and one preserves the setting up where the median of the residues is the weakest. If one of the methods has given a satisfactory set-up, one eliminates all points whose residual is greater than three times the median.

- If the number of exact points exceeds appreciably the number of errors, the setting up is valid, and the points whose discrepancy with the setting up is too high are eliminated. Another method is to really use the photogrammetric knowledge. While doing an approached setting up, errors appear. The only limitation is that most aerotriangulation software are not intended to manage thousands of points automatically generated by algorithms. Adapted tools must therefore be set up.
- On terminating filtering, one wishes to generally choose among the remaining points, according to criteria like the distribution or the a priori validity. Indeed, the distribution of points is not absolutely uniform, some zones being over-equipped in comparison to others. In order to limit this phenomenon, and to limit the number of points in the aerotriangulation, it is necessary to choose among the set of filtered points. Criteria to take into account are the equidistribution of points, the a priori validity of these (in general, one can value the validity of points during their calculation), in order to guarantee having some exact points in every inter-images relation.

Figure 2.4.5 (a) Examples of error sources (top: vehicle moving; bottom: areas locally similar).

Figure 2.4.5 (b) Example of error source: textured zone.

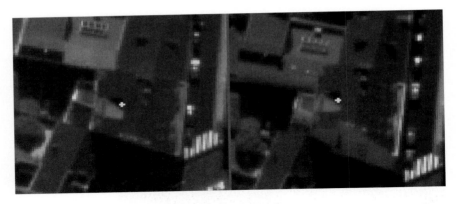

Figure 2.4.5 (c) Example of correct point.

2.4.3.4 Precision

To obtain precise tie points, two main approaches exist. The first consists in attempting to search for sub-pixel positions that maximize the function of similarity. This technique looks mostly like optimization methods. The important point concerning this method is that it links precision and resemblance: the tool used to put the points in correspondence and to localize them precisely is the same.

The second is more geometric, and decouples questions of resemblance and precision. It aims at determining points corresponding to certain details of the ground in a precise way, without trying to put them simultaneously in correspondence (one will nevertheless note that the detection step is not completely independent from the setting in correspondence step; one should indeed preferably choose the points susceptible to being put easily into correspondence, namely those points with good geometric behaviour, like corners, crossings, etc.). A possible way consists in using the theoretical models of corners or positioning points on intersections of contours determined precisely. The putting into correspondence of these points will not change their localization.

Generally speaking, the precision of localization of measure in images depends on the algorithm used for the detection of tie points. The impact of points of interest having a sub-pixel precision on the quality of the aerotriangulation is still, to a very large measure, to be proven. The matching aspect in this type of problem remains also very difficult. Subpixel coordinates of the homologous points can be estimated separately, or jointly during the phase of matching.

For this approach, one can mention the techniques based on theoretical models of corners, or Blaszka and Deriche (1994), or Flandin (1999).

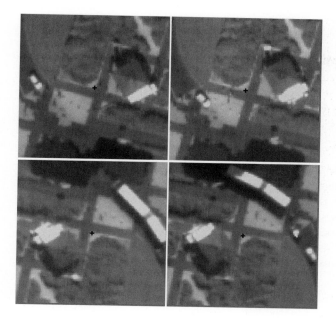

Figure 2.4.6 Example of multiple point of 4th order.

2.4.3.5 Use of photogrammetry

Even though the problem stands upstream of aerotriangulation, photogrammetry can constitute a considerable help in solving the problem of the measurement of tie points.

Photogrammetry can guide methods (by a multi-scale approach) to reduce the combinatories of the problem (by the prediction of homologous positions at the time of the matching), and finally to filter the tie points after their calculation, which contributes appreciably to their reliability.

2.4.4 Discussion

The sections above have exposed some general considerations on objectives and methods bound to the automatic measurement of tie points. We have also to discuss limits and extensions of the existing processes. Indeed, even though the existing systems give good results, there are many problematic cases, as well as numerous possible extensions to less classic cases. These problems may be analysed according to several axes intervening in photogrammetric cases: the sensor, the scene, the image acquisition.

Finally, we will present some algorithmic considerations.

2.4.4.1 The sensor

Use of colour

A priori, the use of colour should allow for a better identification of ground details, but currently it proves to be that most methods use mainly images in grey levels, for the detection of points of interest and for the calculation of tie points as well.

Quality of the signal

One expects a priori to get tie points of better quality with a sensor of better properties: detectors of points of interest generally use the differential properties of the signal; they are therefore sensitive to the image noise. Thus, a better signal to noise ratio in images draws a better stability from these operators. In practice, the evaluation of the sensitivity to the sensor requires the use of important equipment, so as to perform flights in identical conditions, so that no survey has been conducted on the question up to now.

Multi-sensor data

The integration of multi-sensor data is a problem that will occur, because one can consider embarking several cameras simultaneously for a flight (for example, simultaneous embarking of sensors specialized by channel: red, green, blue, and even infrared). No survey on this question and no software dealing with this problem in an operational condition is currently available.

2.4.4.2 The scene

Variation of the aspect of objects

The aspect of objects can vary according to the point of view. In the case of a slope, the perspective distortions can be appreciable (the worse case is facades in vertical aim). The occlusions are not the same. Finally, the aspect can change according to the point of view. Figure 2.4.7 presents two images of the same scene, obtained during the same flight, from two different viewpoints. The superposition of the two images shows distinctly the variations of intensity observed from these two points of view.

Forests, shades, textures

These objects generally create problems in image analysis. In the case of the measure of tie points, they provoke three types of problems. Trees

Figure 2.4.7 Example of change of aspect according to the point of view.

cause problems of precision of pointing. Shades move between two image acquisitions, notably between two strips (the interval of time between two neighbouring images of two different strips can exceed 1 h), and therefore, even though extremities of shades can seem easy to delineate, they are not valid from a photogrammetric point of view (see Figure 2.4.8). Textures naturally lead to problems of identification: it is easy to confuse two very alike details in zones with repetitive motives, typically: passages for pedestrians.

Relief

The relief provokes notable changes of scale and variations of disparity. When looking for homologous points, the space of research is generally larger. In this case, the utilization of photogrammetry on the under-sampled images (multi-scale approach) can prove to be useful.

Case of terrestrial photogrammetry

In this case, often the hypothesis of a plane land (and horizontal) is no longer valid. The space of disparities is more complex than in the aerial case (in the aerial case, with little relief, two images of the same scene are indeed nearly superimposable, which is not the case in terrestrial imagery). Figure 2.4.9 presents the case of a terrestrial scene where a street axis is parallel to the optic axis: distortions bound to the perspective are very significant, as well as the disparities of points.

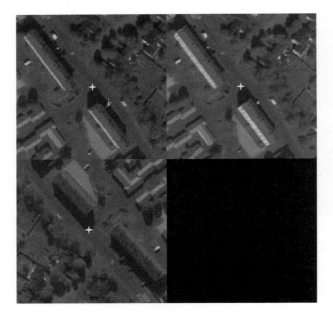

Figure 2.4.8 Example of shade point extremity.

Figure 2.4.9 Terrestrial image acquisition example: the perspective distortions can be significant.

2.4.4.3 The image acquisition

Rotation of images

The interest in using images of any rotation is to achieve some diagonal strips in photogrammetric blocks, which makes them more rigid. Otherwise, if one tries to calculate two overlapping photogrammetric studies simultaneously, the case of large rotations may occur. Techniques of matching generally use linear correlation coefficients between stationary windows.

In the case of known rotation between images, these can be sampled so that these techniques work. In the case of unknown angle rotation, it is necessary to adapt the methods. Invariants by rotation resemblance measures have been developed: one can consider turning the imagettes (Hsieh *et al.*, 1997), or to consider measures relying directly on invariants (Schmid, 1996).

Image acquisitions at various scales

In the case of image acquisitions of different scales, measures of resemblance must also be adapted, because they don't generally support strong changes of scale. One however knows the scales of image acquisitions, which reduces the problem. But even at known relative scales, the problem is significant, because in order to compare two imagettes of two different scales, it is necessary to adapt one of the signals. Otherwise, for example for the positioning of points of interest, errors committed by the detectors may be different between two scales.

Image acquisitions at different dates

If one wishes to process simultaneously some images taken at different dates, it is necessary to be careful after the inversions of contrast that may occur (notably by reason of different positions of the shades). In the same way, the aspect of the vegetation can change. Considering the evolution of the landscape, a certain number of topographic objects can undergo an important radiometric evolution (ageing of roofs).

2.4.4.4 Algorithmic considerations

Parametrage of methods

As in all processes of image analysis, the survey of the parameters must be made in order to learn how to master the method and to know its limits. One will for example aim at reducing the number of critical parameters: one can easily replace a threshold level of a correlation coefficient (that is generally very sensitive) by the introduction of the symmetrical crossed correlation (see description of homologous points in §2.4.1.2).

Number of useful points by image

Studies have shown that there was no improvement to the result of aerotriangulation by using more than 20 points by image (if the 20 points are correct). In the case of automatic measure, it is necessary to have more points to be able to reject reliably the aberrant points. The necessary point number is therefore higher (see §2.4.2.3).

Evaluation

The assessment of the quality of tie points is based on the quality of the aerotriangulation resulting from this setting up. A first criterion of quality is the values of the residues (in microns or in pixels) of tie points. Another technique consists in assigning values to the quality of tie points with the help of reference ground points. These ground points can be calculated with the help of a reliable aerotriangulation or with ground measures.

Combinative

At the time of the research of homologous points between two images, one may a priori attempt to put in correspondence any point of the first image, with any point of the second image. In this case, the combinative of the problem is important. In the case of n images, the combinative increases again. The combinative can be reduced strongly while restricting correctly the space of research of points homologous of a given point.

A classical technique for it is the multi-scale approach: by doing first some calculations at a reduced scale, one may predict the position of a point to search for.

Another solution resides in the 'geometric' approach: if, for a given research zone, one does consider as good candidates only the positions of points of interest that are present, and not all pixels of the region, then the combinative falls appreciably. However, this method rests entirely on the repeatability of the detector of points of interest.

2.4.5 Automatic constitution of an index map

The objective of the constitution of an index map is to calculate 2D information of position and orientation of images of the flight, permitting the placing of all these images in the same reference mark of the plan, so that the homologous details of several images are at the same position in this common reference frame. For two images possessing an overlap, it is then possible to appraise a change of reference frame permitting the forecasting of the position of a homologous point to a given point p. Thus, the index map, beyond its own utility, serves as a guide for the detection of multiple tie points.

Images are treated after a strong sub-sampling. One considers a global reference frame to the plane, common to all images. It is the reference of the block. What one searches for is, for every image I, a similitude permitting the passage of coordinates of a point of the plane (x_B, y_B) expressed in the reference of the block, to its position $p(x, y)$ in the image. The problem is treated strip by strip, then by an assembly of strips. Techniques rely on the detection of points of interest in images, and on the calculation of the relative 2D orientation of a couple of images whose relative rotation is an angle close to 0.

Detection of points of interest

This part of the method includes two parameters: the standard deviation of the Gaussian used for smoothing and derivations, and the size of the calculation window associated to these processes.

Processing a couple of neighbouring images

One supposes that neighbouring images possess a relative rotation of angle of nearly 0, and that because of the strong sub-sampling, the disparities of homologous points of the two images are all similar. Two homologous points can be put therefore in correspondence without the problem of rotation. For two points of interest P_i of the image I, and Q_j of the image J one considers c_{ij} the coefficient of calculated linear correlation on a square window centred on each point. Two points P_i and Q_j are judged homologous candidates if for P_i no point of J is more alike, and reciprocally. Formally:

$$\forall j', c_{ij'} \leq c_{ij}, \quad \forall i', c_{i'j} \leq c_{ij} . \tag{2.15}$$

One thus gets couples of points candidates for the images I and J. In practice, many aberrations are present. One operates therefore a filtering in the space of disparities. For two homologous candidate points, one defines the vector of disparity joining these two points. This is the vector obtained by difference of coordinates of points. While accumulating these disparity vectors in a histogram, one gets a cloud corresponding to the points of similar disparities. These points constitute the homologous points retained by the method.

In practice, the cloud is identified by a morphological closing with an square structuring 3×3 element. After closing, the connex component of the strongest weight is identified, then the points participating in this component are selected. At this step one defines the performance of an image couple: it is the ratio between the number of points participating in the cloud of points and the number of homologous candidates points.

From these points one appraises with the least squares the likeness allowing the best joining of the two references of the images. It permits the second image to be put in the reference frame of the first. This part of the method includes only one parameter concerning the correlation: it is the size of the calculation window. The size of the structuring element for filtering is otherwise stationary.

Process of a strip

One supposes one knows N images constituting a strip. The previous process is applied to the $N-1$ successive couples constituting a strip. One can thus 'sink' all images of a strip in the reference of the first image of this strip. All strips are treated independently. There is no supplementary parameter concerning this step.

Process of a block

One supposes one knows the order of strips. One also supposes that strips possess a correct lateral overlap (> 10 per cent). This part of the method is achieved by iterations: one glues the strip $n + 1$ to the organized block of the n previous strips. One considers therefore that one already has a block organized to which one adds a supplementary strip. It is achieved while first hanging an image of the strip to the block, then hanging up the other images of the strip, image by image.

- To hang up an image of the strip to the block: for an image I of the strip, one temporarily considers all couples between the images I and J of the last strip of the block. One also considers couples between the images I and J having undergone a rotation of π. For all formed couples, one performs the process of image couples studied earlier, then one keeps the couple possessing the best performance. If this performance is not sufficient, one repeats the operation until one finds an effective enough couple. One thus gets a couple in an image I_0 and J_0 permitting the image I_0 to be put in the reference frame of the block.
- Progressively one sinks images of the strip in the reference of the couple. To avoid error accumulations, an optimization of the system is done at regular intervals (every n_{opt} images), by a method described in the following.

This part of the method introduces two parameters. The first is the threshold of performance beyond which one accepts the coupling of image between the strip and the block. This threshold is fixed to a value that a precise survey showed was quite reliable. The second is $n_{opt} = 5$. This value allows the method to be reliable while reducing the number of optimizations.

Optimization of block

This optimization aims at refining the positions and orientations of the images of a block. This is necessary because it is not possible of 'to glue' in a rigid way two strips because of distortions that accumulate at the time of the constitution of the intra-strip index map. One considers all couples of images possessing an overlap. For each couple, the homologous points to the previous sense are calculated. A function of cost aiming to achieve the objective of the image assembly is minimized. Technically this function is the sum of cost functions bound to every couple of images. The function bound to a couple of images aims to minimize, for each of the homologous points of the image couple, the distance of the two homologous in the reference of the assemblage image. This function is adjusted so as to avoid the aberrant points.

Survey of parameters for the constitution of the index map

The behaviour of parameters is known. It appears notably that for some among them, values are stationary in the sense where meaningful studies have been processed without modification. These are:

- the value of the standard deviation of the Gaussian in the calculation of points of interest, as well as the size of the associated calculation window;
- the size of the correlation window for the calculation of homologous candidates points;
- the size of the structuring element used in filtering in the space of disparities.

For the other parameters, their behaviour is known: their variation allows an increase in reliability at the cost of more calculation time in any case, or inversely. In this sense, the sensibility of the parameters is low as their influence on the results is identified.

Figure 2.4.10 Excerpt of an index map (city of Rennes, digital camera of IGN-F). 14 strips of 60 to 65 images. The geometric precision of mosaicking is good: the errors on the links range to 1 or 2 pixels.

2.4.6 Conclusion

The problem remains wide open. Nevertheless, under certain conditions, some systems are quite capable of producing tie points adapted to calculate photogrammetric works. These tools can be very useful for large studies.

In any case, digital photogrammetry will need such tools, which justifies completely the ongoing developments in this domain: the problematic zones (occlusions, strong relief, texture, change of date, strong scale variation, etc.) remain difficult points for these methods.

References

Blaszka T., Deriche R. (1994) Recovering and characterizing image features using an efficient model-based approach. *Research Report from INRIA*, Roquencourt, France, no. 2422.

Flandin G. (1999) Détection de points d'intérêt sub-pixellaires pour l'orientation relative 3D de images. *Stage Report*, Institut Géographique National France.

Förstner W., Gülch E. (1987) A fast operator for detection and precise location of distinct point, corners and centre of circular features. *Proceedings of the Intercommission of the International Society for Photogrammetry and Remote Sensing*, Interlaken, Switzerland, pp. 281–305.

Harris C., Stephens M. (1988) A combined edge and corner detector. *Proceedings of the 4th Alvey Conference*, Manchester, pp. 147–151.

Hartley R.I. (1997) In defense of the Eight-Point Algorithm. *IEEE Transactions on pattern analysis and machine intelligence*, vol. 19, no. 6, June, pp. 580–593.

Heipke C. (1998) Performance of tie-point extraction in automatic aerial triangulation. *OEEPE Official publication*, no. 35, pp. 125–185.

Heipke C. (1999) Automatic aerial triangulation: results of the OEEPE-ISPRS test and current developments. *Proceedings of Photogrammetric Week*, Institute for Photogrammetry, Stuttgart, pp. 177–191.

Hsieh J.W., Liao H.Y.M., Fan K.C., Ko M.T., Hung Y.P. (1997) Image registration using a new edge based approach. *Computer Vision and Image Understanding*, vol. 67, pp. 112–130.

Schmid C. (1996) Appariement d'images par invariants locaux de niveaux de gris. *Thèse de doctorat de l'institut National Polytechnique de Grenoble*.

Zhengyou Zhang, Gang Xu (1996) *Epipolar Geometry in Stereo, Motion and Object Recognition*. Kluwer Academic Publishers, pp. 102–105.

2.5 DIGITAL PHOTOGRAMMETRIC WORKSTATIONS

Raphaële Heno, Yves Egels

2.5.1 Introduction: the functions of a photogrammetric restitution device

Whatever its principle of realization, one will find in any photogrammetric restitutor a certain number of basic functions that one can represent on a working diagram (Figure 2.5.1). This principle of working was discovered during the first industrial realizations at the beginning of the twentieth century (restitutor of Von Orel at Zeiss, Autograph of Wild). In the first systems, the geometric function is achieved by an optic or mechanical analogue computer (whence their 'analogical restitutors' name). The two other functions use a human operator, whose stereoscopic vision assures the image matching, and whose cultural knowledge the interpretation.

In the 1970s, a first appreciable evolution appeared, the analytic restitutor: image formulas are calculated analytically by a computer, which displaces images from an optic system using a servo, because images being

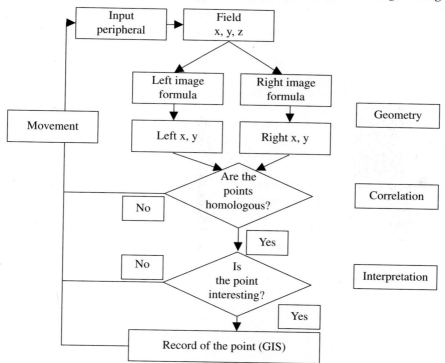

Figure 2.5.1 Schematic diagram of the photogrammetric restitution.

on films, the matching as well as the interpretation are always done by a human operator.

Today, images become digital, and can be visualized on a simple computer screen. Besides, non-geometric functions are more often taken into account by algorithmic means, and the human operator, and therefore the visualization itself, then becomes superfluous. In this last case, there isn't a photogrammetric restitutor strictly speaking, but only a specialized software functioning on more or less powerful computers.

In the following, one is concerned mainly with the case where a human operator must, instead of achieving the totality of photogrammetric process by hand, at least supervise it, control its results, and possibly perform certain operations by hand.

2.5.2 The technology

2.5.2.1 *The equipment*

The passage to the digital in photogrammetry allowed a freedom from the high-precision optic and mechanical components, which kept the acquisition and maintenance costs of systems at a very high level. In addition, the size of the machines was considerably reduced (see Figure 2.5.2).

Systems of digital photogrammetry are based on workstations or, increasingly, on PCs. The recent computers include almost the whole of the necessary means for photogrammetric restitution, and today have enough power for these applications. Nevertheless it is necessary not to forget that the manipulated data images generally have sizes of several hundreds of megabytes, and that a given study will require the use of tens, or even hundreds of images. The capacity of storage, and access times, should therefore be especially adapted.

Only two indispensable elements are not yet completely standard, even if the development of video games makes them more and more frequent: the peripheral for data input and the stereoscopic visualization; as regards the computer conception, digital photogrammetry is hardly more than a video game with a technical aim.

Peripheral for data input

The system must possess a peripheral of coordinate input able to address three independent axes. Solutions are numerous, from arrows on the keyboard, to the cranks and pedal of traditional photogrammetric devices, or controllers of immersion of virtual-reality stations, joysticks, or the specialized mouse. Only the cost/ergonomics ratio can overrule them. The practice establishes that the movement must be as continuous as possible, and that the control in position is far preferable to speed commands (joysticks). When a specialized peripheral is developed, supplementary buttons may be usefully attached, allowing one to simplify the use of the most frequent commands.

Stereoscopic visualization

The system must also allow stereoscopic vision, inasmuch as one will ask the operator to achieve some visual image matching. Here also, several solutions exist, and some are under normalization. It is about presenting to each eye the image he has to observe. In order of increasing comfort, one can mention:

- the display of the two images in two separated windows, which the operator will observe with an adequate optic system of stereoscopic type, placed before the screen;
- the simultaneous display of the two images, one in green, the other in red, the operator benefiting from the structural complementarity of such glasses (anaglyphs) whose cost is extremely low;
- the alternative display of the two images, a device using liquid crystals letting the image pass toward the eye to which it is destined.

In the professional systems, only this last solution is used, due to its better ergonomics. The frequency of the screen conditions the quality of the stereoscopic visualization, because the alternative display divides by two the frequency really discerned by the user (a minimal frequency of 120 Hz is necessary to avoid operator fatigue). Several convenient realizations are susceptible of being used. The liquid crystals may be placed directly before the eye, a low-cost solution, synchronized by a wire or by infrared link. Or they may be placed on the screen, the operator then using a couple of passive polarized glasses (which allows him or her to look indifferently

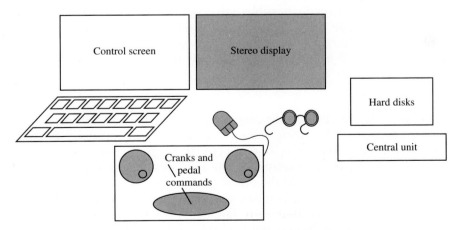

Figure 2.5.2 Diagram of a digital photogrammetry system.

at the stereoscopic screen and at the command screen), the most comfortable solution but also the most expensive.

2.5.2.2 Photogrammetric algorithms

It was not necessary to reinvent photogrammetry to make it digital. The equations are the same as those on the analytic photogrammetry systems (equations of collinearity or coplanarity).

More and more systems are 'multi-sensor', that is to say that they can not only process aerial images, but also different geometry images, for example the images from scanning sensors or from radar. The setting up of these images uses mathematical models different from the traditional photographic perspective, possibly parametrable by the user.

Whatever their origin, once images are set up or georeferenced, the function 'terrain → image' allows one to transmit in real time the displacement terrain introduced with the mouse or the cranks.

On the other hand, contrary to the analogical or analytic restitutors, the availability of the image under digital shape allows a considerable spread of the possibilities of the automation of photogrammetry, requiring to bring together in one unique system photogrammetric functions and functions of image processing and shape recognition.

2.5.3 The display of image and vector data

2.5.3.1 Display of images

The used images always have some sizes much greater than the dimension of the screen (often 100 times larger). It will thus be necessary to be able to displace the display window conveniently within the total image. Two modes of displacement are possible: either the image is fixed, and the cursor of measure mobile; or the cursor is fixed and central, and the image is mobile. The measure will be performed by superimposing a pointing index (whose colour will be adjusted automatically to the background image).

The image is fixed

The fixed image/mobile cursor configuration is easiest to implement: it is just a question of displacing some octets in the video memory. But the ergonomics of this solution is mediocre. If images are not reprocessed geometrically (epipolar resampling), the transverse parallax is eliminated in only one point, generally the centre of the image. During the displacements of the cursor, some parallax appears, which makes the pointing if not impossible, at least imprecise. Besides, when the cursor reaches the side of the screen, the system must reload another zone of images, which interrupts the work of restitution and distracts the operator.

The cursor is fixed

More demanding in calculation capacities, the mobile image/central cursor configuration is far more preferable. If the displacement of images is fluid enough, and it must be so not to tire the operator, this configuration recalls the analytic devices, where image-holders move in a continuous way in front of a fixed mark. Images are generally charged in memory by tiles, to optimize the time of reloading.

The system sometimes proposes a global view (under-sampled image) for the fast displacements in the model.

In the 'exploitation of the model' mode, it is preferable to dedicate the maximum surface of the stereo screen to the display of the two images. But then, at the time of aerotriangulation points measurement, it is convenient to be able to display for every point all images where it is present (multi-window).

2.5.3.2 Zooms and subpixel displacement

Zooms by bilinear interpolation, or to the nearest neighbour, serve to enlarge all or part of the current screen. Their goal is usually to allow to point with a better precision than the real pixel of the image. Unfortunately, the zoom decreases the field of observation simultaneously and adds a significant fuzziness to the image, these two effects degrading appreciably the quality of the interpretation. Subpixel displacement of the image allows these two defects to be palliated simultaneously (but this function is not generally available in the standard graphic libraries, and requires a specific programming).

In the case where the image is fixed, it is enough to resample the cursor so that it appears positioned between two pixels, which requires that it is

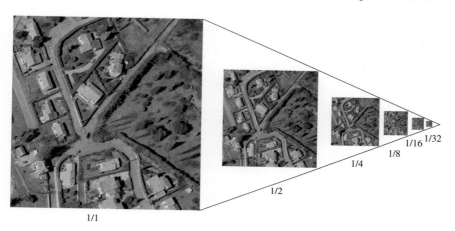

Figure 2.5.3 Rear zoom pyramid.

even-formed of several pixels. If the image is mobile, then it is the totality of the displayed image that it is necessary to recompute.

Concerning the rear zooms, they are immediately calculated or generated in advance under the shape of a pyramid of physical images of decreasing resolution (see Figure 2.5.3). In the case of image pyramids, one won't be able to zoom back at any scale.

2.5.3.3 Image processing

Within basic tools one finds at least the functionalities of adjustment of the contrast, the brightness, of positive/negative inversion. Based on the colour table (LUT, a term that stands for look-up table) of images, they work in real time, and don't require the creation of new images.

It is sometimes possible to apply convolution filters (contours improvement, smoothing ...). Their application being quite demanding for the CPU, it slows down considerably the time required for image loading. Besides, these filters often have some perverse effects on the geometric plan (displacement of the contours), and the influence on the precision of the survey can be catastrophic. For a better efficiency, one will prefer sometimes to calculate new images.

2.5.3.4 Epipolar resampling

The visualization of a stereoscopic model whose images have been acquired with appreciably different angles can be tedious for operators, because the two images have too different scales and orientations. Performances of usual image-matching techniques are degraded also by this type of images. (See Figure 2.5.4.)

In this case, it is desirable to make an epipolar resampling: one calculates the homographic plane that allows one to pass from the initial image

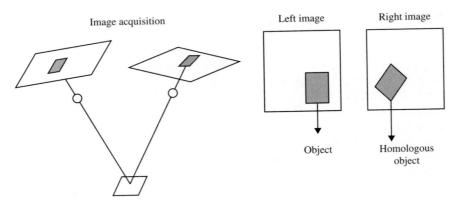

Figure 2.5.4 Visualization of stereoscopic model using images with significant differential rotations.

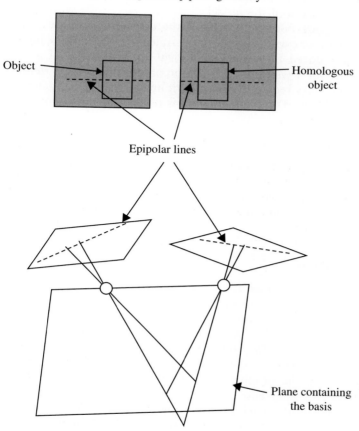

Figure 2.5.5 Epipolar resampling.

to the one that would have been obtained if optical axes of photos had been parallel between them and perpendicular to the basis (normal case); after this process, two points of a line of the left image will have their counterparts in one line of the right image, and the transverse parallax is constant. Epipolar lines are the intersections of the bundle of planes containing the basis and of the two image planes. In the resampled images, these lines are parallel, whereas they were converging in the initial images. (See Figure 2.5.5.)

This resampling requires at least the calculation of the relative orientation (and of course cannot be used at the same time as the measures necessary for setting it up). It generates two new images, which must be taken into account in the management of the disk space. As it is valid for only one couple, it is possible to cut up images and to keep only the common parts of images of the model.

Epipolar resampling not only improves the operator's comfort, but it accelerates the work of correlators, since the zone of research of the homologous points is reduced to a one-dimensional space (instead of a two-dimensional space for a systematic correlation of images). Some systems of digital photogrammetry require working with images in epipolar geometry for the extraction of the DTM; others do this resampling continuously.

2.5.3.5 Display of the vectors

Whereas on the analytic restitutors the addition of a vector display on the images (very useful for cartographic updating) was a very costly option, requiring complex optical adaptations, the superposition of vectors in colour to the images in stereo doesn't pose any problem for digital photogrammetry systems.

2.5.4 Functionalities

Digital restitution systems offer at least the same functionalities as the analytic restitutors. Besides, these can practically be reused without change, except for the addition of functions to help pointing, which profit from the digital nature of the image. But their features allow one to consider numerous extensions of their use. We summarize here the main among them.

Contrary to what happened in analogical or analytic systems, the geometric quality of the products calculated on the digital photogrammetry system (vector data base, DTM, orthophotos) will be reasonably independent of the system. The geometric algorithms used are very close, and no mechanics intervenes. The only difference lies in the pointing capability which may be at the pixel level, or using sub-pixel methods. One should especially be concerned with the quality of the following steps:

- condition of the films;
- quality of the digitization (type of scanner used, control of the parameters, scanner resolution);
- quality of the control points;
- quality of measures of points of relative and absolute orientation.

2.5.4.1 The management of data

The manipulated images

Digital photogrammetry workstations are capable of working with images originating from scanned classic aerial pictures, scanned terrestrial images, or with the images of aerial or spatial digital cameras (e.g. the CCD matrix digital camera of the IGN-F, the DMC of Zeiss Intergraph, the APS 40 of Leica Helava Systems, see §1.5).

The black and white 8 bits and colour 24 bits images are easily read by digital photogrammetry systems. Considering that the images may have 12 bits or more by channel that are meaningful for the digital sensors, it happens that only the first 8 bits of every pixel are used. Anyway, the current computer screens actually display 6 bits only, which is also the true dynamics of movies, and the human eye only distinguishes a part of the 256 levels of grey of a coded image on 8 bits, but with the large dynamics digital sensors, this loss of information is prejudicial for algorithms of image matching and automatic shape recognition.

There is no consensus within manufacturers of digital photogrammetry systems on the computer format of images to use. Nearly all are capable of reading and writing the TIF (tiled or not), but some recommend converting it to a proprietary format to optimize process times. It is the same for the JPEG compression: its direct utilization slows down some applications, as this technique (in its present state) prevents the direct access to portions of image. The wavelet compression gives excellent results in terms of the ratio 'quality of compressed image/reduction of volume', and it is likely that the digital photogrammetry systems will quickly adopt this process.

Photogrammetric database

The data necessary to important studies exploitation are quite numerous and voluminous; they require a rigorous organization that is taken into account, either by a hierarchy predefined by directories, or by a specific database.

Systems generally allow one to store data by project, that correspond intuitively to a given geographical zone, to a theme, etc. A project can correspond to a directory of the arborescence of the system, in which one finds the files necessary for setting up the models (files of camera, files containing reference points, possibly files containing the measures on the image achieved on another device . . .), as well as files generated by calculations (parameters of the internal orientation, position and attitudes of the camera for each of the cliches, matrixes of rotation . . .). Data are managed by model, or image by image (a file by model, or a file by image). Images are stored in this same directory, or, for convenience in the management of the disk's space, merely referenced there. (See Figure 2.5.6.)

The local networks (Ethernet 100, then Giga Ethernet) allow work in architecture on a customer-server basis: images are set up and stored on a machine 'server', while machines' 'customers' use them (visualization, process, survey) via the network, which requires mechanisms of control of the data consistency.

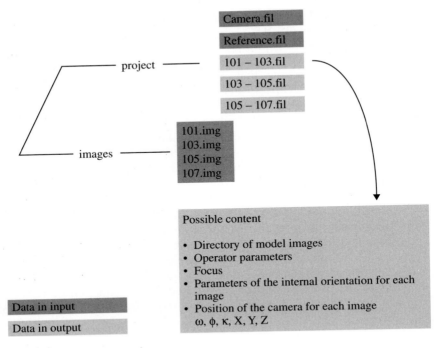

Figure 2.5.6 Example of photogrammetric database.

2.5.4.2 Basic photogrammetric functionalities

Setting up of a stereoscopic couple

INTERNAL ORIENTATION

The internal orientation phase is always necessary for images coming from a digitized classic aerial photograph. Knowledge of the theoretical position of reference marks on the bottom plate of the camera (via the certificate of calibration of the camera) and the shape of these reference marks allowed at least a partial automation of this phase of measures, thanks to image matching.

The operator is sometimes asked to search by hand for the first reference marks, the system recognizing the next ones automatically. Automatic image matching, if necessary, assists interactive pointing.

The internal orientation is completely automatic on certain systems.

EXTERNAL ORIENTATION

Most systems calculate the relative and absolute orientations by the simultaneous use of the collinearity equation. Even though the measure of points of relative orientation is assisted by image matching, or even completely automatic in cases where one uses a theoretical distribution of points, it is always imperative that an operator marks by hand the position of reference points.

The elimination of the parallax in X and in Y is generally performed by blocking one of the two images, and by moving the other with the positioning system (cranks, mouse). Once the choice of the fixed image is made, one can ask for a pointing by image matching. In certain cases, this one can fail (vegetation, periodic textures, homogeneous zones); the use of the image matching does not dispense with a pointing by the operator, even if it is not so precise, because it remains more reliable in spite of everything.

As soon as enough points are measured (six points to have residues of relative orientation, three ground points), one can launch a first calculation. Then the function 'ground → image' is known, and allows one to move automatically, for example on a supplementary control point. When measures are sufficiently overabundant, the examination of calculation residues allows one to control the quality of the setting up, and to validate it.

Measures and acquisition in a stereoscopic couple

The acquisition is very often limited to a stereoscopic couple; the automatic loading of images of the neighbouring couple on the edge of the image is sometimes possible, but of course requires having all images concerned already loaded on the device.

Basic functionalities of the digital photogrammetric system are:

- acquisition of points, lines, surfaces;
- destruction, edition of these structures.

Ideally, the digital photogrammetric system is interfaced with a GIS that manages data acquisitions in a database.

All functionalities of the GIS (spatial analyses, requests, multi-scale cartography) are then directly usable after the data acquisition. Data are structured according to a more elaborate topology (line, spaghetti, sharing of geometry, object structure).

Data extracted from the digital restitutors are systematically tridimensional. But rare are yet the GIS that allow one to manage completely a 'true 3D' topology (management of vertical faces, of overhangs, of clearings), and not only of the 2.5D (planar topology, only one Z attribute by point).

The GIS generally propose interfaces toward the most current formats (DXF, DGN, Edigeo ...).

2.5.4.3 Advanced photogrammetric functionalities

Aerotriangulation

Aerotriangulation allows one to georeference simultaneously all images of a site, using as much as possible the overlaps that they have in common, with a minimum number of reference points. This operation starts with a phase of measure of the image coordinates of a certain number of points seen on the largest number possible of images. Then a calculation in block allows one to determine the set of photogrammetric parameters of the work. The modules of available aerotriangulation calculation on systems of digital photogrammetry use the same formulas as the analytic aerotriangulation, and are most of the time identical.

The measure of reference points, always interactive, is helped by the multi-windowing: once a point is measured on a image, the system can display in mini-windows all images susceptible of containing it, directly zoomed on the zone concerned. This is made possible by the a priori knowledge of the overlaps between images and their position in relation to each other, determined by GPS, or via the assembly table defined by the operator. The only work remaining for him is to measure the point position in the window in which it is present, in monoscopy or in stereoscopy as well, or with the assistance of an automatic image matching process.

On another hand the measure of tie points is very automated: one can replace the precise measurement done by a human operator by automatic image-matching measures in very large number, either by the operator's choice helped by the multi-windowing, or by automatic selection of tie points by an adequate algorithm. After filtering the wrong points, enough points normally remain to assure the stability of the block.

Altimetry and digital terrain models (DTM)

The DTM can be used either as supporting data to the restitution, or as a product. In addition, the digital photogrammetric workstations are perfectly adapted to the control and the correction of DTM obtained by automatic methods.

In the case where a DTM pre-exists (result of a previous restitution, of a conversion of level lines ...), it is possible to fix the measure mark to the DTM, which frees the operator from a tedious task. However, this help is quite limited, not only because of the frequently excessive generalization of the available models, and of the non-representation of the objects over the ground.

In an intermediate way, it is possible to help the restitutor thanks to a real-time correlator, which replaces altimetric pointing. One uses in this case a correlator working in the object space, fixing the altitude to the maximum of the image matching detected on the vertical determined by the system of command. But the correlator must be considered as a help, and up to now it may work only in a supervised mode: many false correlations may happen, especially when the signal/noise ratio is low, or for example on specific periodic structures.

The complete make-up of a DTM is usually let to a non-interactive process, possibly guided by some initial manual measures; indeed, the manual acquisition of a DTM is a long and very trying operation. One prefers to limit the human interventions to the control and the correction. The control is often performed visually, essentially by stereoscopic observation of level lines calculated from the DTM superimposed onto the couple of images. For corrections, each worker imagined some different solutions, from the local recalculation with different parameters, with possible integration of complementary human pointings, up to more brutal operations on altitudes themselves: new acquisition, levelling, digging, interpolation.

Aids to the interpretation

The functionalities of automatic or at least semi-automatic data extraction are awaited with impatience by users, who invested in systems of digital photogrammetry while hoping that algorithms of researchers would allow them to quickly improve their efficiency. But with the exception of automatic extraction of the DTM, and the cartographic restitution of contour lines, no tool is yet industrially implemented on systems of digital photogrammetry.

Orthophotographies, perspective views

Digital photogrammetric stations are often seen as machines to manufacture orthophotos. But actually, with the exception of the preparatory phases (reference points of the aerotriangulation, control and correction of the DTM) presented in the previous paragraph, it would be more economic to use a specialized software functioning on a standard computer.

2.5.4.4 Limitations

Photogrammetric workstations allow one, fundamentally, to perform at least the same work as analytic restitution devices, which they are progressively going to replace. Nevertheless, it is certain that the ocular comfort of operators is not equivalent to that of their predecessors, far from it. But as these materials directly follow the possibilities offered by the PC for video games, where the problem is quite similar, one can expect improvements of visual comfort, a point that employers are more and more obliged to take into account.

In addition, these stations allow processes otherwise impossible to make on the same workstation, such as the realization of orthophotographies.

Let us note finally that these stations obviously require images under digital shape, that are provided at present especially by digitization of argentic images, which implies a phase of quite expensive digitalization work currently, and a significant supplementary delay. The generalization of the use of digital cameras providing directly digital aerial images will make this situation evolve. In the intermediate situation that prevails to the date of writing of this work, it is essentially the price of acquisition and maintenance of equipment (that are hardly more powerful than simple PCs) and of software (that democratize themselves considerably) that make these stations attractive.

3 Generation of digital terrain and surface models

INTRODUCTION

The mathematic modelling of a landscape is a very important step of the photogrammetric processes. It has several key applications, and among them the cartography of the intervisibility (use of digital models for telecommunications, cellular phones for example), the hydrologic studies, the preparation of cartographic features (the contour lines), and the preparation of orthophotographies. Thus we shall review some important points that are not, strictly speaking, bound to the digital aspect of the photogrammetry itself, but that in fact are such a substantial output of today's photogrammetry that it is necessary to analyse them here as completely as possible. We will see how are defined and specified the various surface models (§§3.1 and 3.2), how the data samples are produced (§3.3), how to remove raised structures when one wants to get a DTM (§3.4), how to build an optimal triangulation of a digital model so as to describe a surface in the best way possible (§3.5), how to extract automatically the characteristic terrain lines, as these lines are necessary to minimize unpleasant artefacts of any DTM, particularly for hydrologic matters (§3.6). Then we will conclude by some definitions, some quality considerations and some practical remarks about the production of digital orthophotographies (§§3.7, 3.8 and 3.9).

3.1 OVERVIEW OF DIGITAL SURFACE MODELS

Nicolas Paparoditis, Laurent Polidori

3.1.1 DSM, DEM, DTM definitions

A digital elevation model (DEM) is a digital and mathematical representation of an existing or virtual object and its environment, e.g. terrain undulations within a selected area. DEM is a generic concept that may refer to elevation of ground but also to any layer above the ground such as canopy or buildings. When the information is limited to ground elevation, the DEM is called a digital terrain model (DTM) and provides information about the elevation of any point on ground or water surface. When the information contains the highest elevation of each point, coming from ground or above ground area, the DEM is called the digital surface model or DSM.

Natural landscapes are too complex to be analytically modelled, so that the information is most often made of samples. Theoretically, a genuine 'model' should also include an interpolation law that would give access to any elevation value between the samples, but this is generally left to the end user.

Together with the elevation data, the specification of a DEM is provided by ancillary data. The specification, which is a description of the data set, is necessary to let users access, transmit or analyse the data. It consists of a number of characteristics that may be given as requirements.

3.1.2 DEM specification

The specification of a DEM generally includes two kinds of parameters.

On the one hand standard altimetric specifications do not differ from the case of analogue maps, typically geodetic parameters (ellipsoid, projection, elevation origin ...) and geographic location (e.g. coordinates of corners) but not scale, which is meaningless in the case of digital maps.

On the other hand, a DEM is a digital product that cannot lead to an altimetric grid without a few specifications:

- the digital format, i.e. a type (integer, character, real ...) and length (often 2 bytes);
- the significance of numerical values, i.e. unit (metre or foot) and in some cases, the coefficients of a conversion law, for instance a linear transform that makes the values fit a specified interval;
- the grid structure, which may be irregular (e.g. triangular irregular networks or digitized contour lines) or regular (typically a square mesh regular grid);
- the mesh size, which is important in the case of a square mesh regular grid – not to be considered as a resolution.

The impact these specifications may have on the quality of the DEM is discussed in §3.2.

The three main widespread models are regular raster grids, triangular or planar faces irregular networks, and iso-contour and break lines networks.

3.1.3 Digital models representation

3.1.3.1 Regular raster grid (RG)

Regular raster grids are well adapted to the representation of 2.5D surfaces, i.e. surfaces that can be described by a 3D mathematical function of the form $z = f(x, y)$. Regular raster grids describe a regularly sampled representation of this function f which implicitly assumes a definition of a projection (where the form $z = f(x, y)$ is valid) defining the expression of the (x, y) coordinates from the initial 3D coordinates in a global Cartesian reference system. When the surface describes the Earth's relief this projection is a map projection given an ellipsoid.

Indeed, RGs have the geometry of an image where the pixels are the nodes of the regular raster grid and the grey values of the pixels represent the elevations. Indeed, one of their main advantages is that they can be visualized as grey-level images or in colour with a look-up table, e.g. hypsometric. They should also preferably, for data size reasons, be stored as images. To pass from the image to the grid geometry and information, some parameters (x_0, y_0, dx, dy, b) and the units in which these parameters are expressed, in addition to the map projection system and to the ellipsoid parameters, have to be known and stored either in the image/grid header or aside in a separate file. The transformation from the image coordinates of pixel (i, j) to corresponding 3D coordinates (x, y, z) can be expressed as:

$$x = i\,dx + x_0$$
$$y = j\,dy + y_0$$
$$z = G(i, j)\,dz + b. \qquad (3.1)$$

$G(i, j)$ is the grey level of pixel (i, j). (x_0, y_0) are the spatial coordinates of the image's first row and line pixel. (dx, dy, dz) are the spatial sampling of the grid respectively along the x, y and z axes. b is the elevation corresponding to the grey-level 0 in the image. The dz and the b parameters are necessary if the grey levels are not coded as floating values but as 8 or 16 bits integers.

As all sampled models, a raster grid can describe many possible surfaces. This problem arises when we want to determine the elevation of a point falling inside the grid in between the known nodes of the grid. The elevation has to be calculated from the elevations of the neighbouring grid points with an interpolation function (bilinear, bicubic, etc.).

The problem with the RG models is that the spatial sampling of the grid is regular and thus some features of the landscape can be correctly described at a given spatial sampling while some others, smaller, would not be relevantly sampled thus smoothed or even missing in the sampled model. Using the same density of samples all across a changing landscape is definitely a limit of these models. Adaptive sampling meshes are more adapted to describe an irregular world. The distribution of elevation points needs to be dense on rough relief areas and only sparse on smooth areas. Indeed, the points do not need to be dense but to be well chosen to describe the surface as well as the application requires.

Moreover, an underlying hypothesis made in aerial photogrammetry is that we suppose that the surface we are trying to model can be described by a graph of the form $(x, y, f(x, y))$. Indeed, this hypothesis is not always valid, e.g. in urban areas due to 3D discontinuities. It is even less valid in the case of terrestrial photogrammetry where overlapping surfaces often occur. RGs are an easy but not a general way to model surfaces.

3.1.3.2 Triangular irregular networks (TIN)

Most of the conventional data acquisition systems provide sparse point measurements. Building a regular grid DSM from these samples is thus often against nature. The idea of triangular irregular networks is to adapt completely the model to the samples by describing the surface by elementary triangles where the vertices are the samples themselves. These triangles can be constructed from the samples limited to their planimetric components by a 2D Delaunay triangulation process (a very good triangulation algorithm is available on the web site of the Carnegie Mellon University) if the surface is 2.5D and by a more complex 3D Delaunay triangulation process, also called tetraedrization, if not. This triangular modelling is widespread and extremely popular in virtual reality and in CAD world and systems.

The triangles built by a triangulation process have only the aim of defining neighbourhoods in which one can directly calculate the elevation using an interpolation function between the three vertices for a given (X, Y).

The simplest interpolation function is the following. Let $T(P_1, P_2, P_3)$ be the considered triangle where $P_1(X_1, Y_1, Z_1)$, $P_2(X_2, Y_2, Z_2)$, $P_3(X_3, Y_3, Z_3)$ are the three triangle vertices. Let $V((X_1,Y_1,Z_1) - (X,Y,Z))$, $V_1((X_2, Y_2, Z_2) - (X_1, Y_1, Z_1))$, $V_2((X_3, Y_3, Z_3) - (X_1, Y_1, Z_1))$, and (α, β, γ) be the barycentric coordinates of point P inside T. V lies inside the plane defined by (V_1, V_2) if V can be expressed in a unique way under the form $V = \mu V_1 + \nu V_2$ where:

$$\mu = \frac{(V \wedge V_2)k}{(V_1 \wedge V_2)k} \quad \text{and} \quad \nu = \frac{(V \wedge V_1)k}{(V_2 \wedge V_1)k} \quad \text{where} \quad k = \begin{pmatrix} 0 \\ 0 \\ 1 \end{pmatrix}. \quad (3.2)$$

Thus $Z = (\mu V_1 + \nu V_2)k$.

Contrary to RGs, the samples and the set of triangles could be non-ordered, thus leading to more time-consuming elementary operations as interpolating the Z elevation value for a given (X, Y). Indeed all triangles have to be parsed to determine the whole set of triangles including this point. For a given (X, Y) and for each triangle T within the set of triangles we can calculate the coordinates of $P(X, Y, Z)$ belonging to the plane lying on T as described above. Under the assumption that the surface is 2.5D, the parsing of the set of triangles can be stopped when $P(X, Y, Z)$ belongs to T which is verified if and only if:

$$\begin{cases} \lambda = 1-\mu-\nu \geq 0 \\ \mu \geq 0 \\ \nu \geq 0 \, . \end{cases} \quad (3.3)$$

Adding some easy topology (ordering information) by storing for each triangle the index of the three adjacent triangles limits considerably the number of triangles to be parsed. Indeed, let us consider a triangle T_0. We would like to determine the triangle T containing the point (X, Y). Starting from T_0, we will look for the adjacent triangle T_1 of T_0 in the direction of (X, Y). We will start this process again from T_1 and again until the current triangle contains (X, Y). This is also a way of parsing the triangles to generate an elevation profile between two 3D points on the surface.

Some further topology that we will call spatial indexing can accelerate in an impressive way the number of triangles to test. Indeed, the 2D (x, y) space can be regularly and recursively split, e.g. in a dichotomy process, in square bounding boxes and the patches can be sorted in a tree graph where every branching node would represent a rectangular bounding box – which would get smaller while climbing up in the tree – and where the leaves at the end of the branches would be the patches themselves. One should remark that a patch/leaf could belong to several boxes/branches. We here obtain a hybrid mix of an irregular sampling with a regular ordering which keeps at the same time the useful flexibility of adaptive meshes to describe the relief variations, and the rapidity to access to interpolated information.

This spatial indexing can also be given by an index raster map giving for each (X, Y) node of the map grid the index of the corresponding triangle. This map can be constructed by filling each triangle (limited to its planimetric components) with the label/index value of the triangle inside the map grid. The quality of this map will depend on its spatial sampling considering the aliasing problems arising close to the triangle limits.

A drawback of raw TIN models is that the slope is identical on the whole facet surface and the slope is discontinuous between adjacent facets. If we suppose that our sample acquisition technique provides us in addition the surface normal vector and if we assume that the surface is smooth and curved inside the facet and/or across the facets, we can improve the surface approximation by adding to each triangle some more parameters to model locally the behaviour of the surface by an analytical function, e.g. bicubic splines, compatible in continuity and in derivability to all adjacent facets.

3.1.3.3 Surface characteristic lines

A surface can also be described by characteristic feature lines (or points) and by contour lines (also called soft lines) giving the intersection between the surface and planes regularly sampled in one direction. This modelling is valid if the surface is 2.5D. Contour lines can be directly acquired from manual stereo-plotting of a stereopair, through analogue, analytic, or digital devices, derived from a regular grid or a TIN model DSM, or digitized manually or automatically from scanned maps. The reconstruction of a surface from contour lines is well conditioned if the surface is smooth. If not the addition of characteristic breaking lines (also called hard lines), e.g. slope break lines, helps in the regularization of the reconstruction problem by describing the surface local high derivatives.

Modelling a surface in this way can also be seen as a data compression of the true surface. And the density of contour lines can be seen as a data compression rate vs. a data loss ratio.

To conclude, the choice of one of these models and on its parameters will depend on the 3D geometric particularities of the object to model, on the requirements of the application using the model, and will have a definite impact on the surface approximation accuracy and on the data storage size.

3.2 DSM QUALITY: INTERNAL AND EXTERNAL VALIDATION

Laurent Polidori

3.2.1 Basic comments on DSM quality

Many different techniques can be used to extract a DSM, depending on available data, tools or know-how: map digitization and interpolation, optical image stereo correlation, shape from shading, interferometry, laser altimetry, ground survey, etc.

In spite of their variety, each of these techniques can be described as a two-step process:

- the first step consists in computing 3D locations for a large number of terrain points;
- the second step consists in resampling the resulting data in order to fit a particular grid structure and a particular data format.

Therefore, the quality of a DSM is the result of the way these two steps have been carried out.

DSM quality

What is the quality of a digital surface model? Basically, it is its capability to describe the real surface, but quality criteria cannot be defined without keeping in mind user requirements. The impact of DSM quality on the reliability of derived information (such as geomorphology or hydrography) has been analysed by many authors to improve the understanding of what DSM quality means (e.g. Fisher, 1991; Lee et al., 1992; Polidori and Chorowicz, 1993).

The relevance of a criterion depends on the way it reflects these requirements, and on the feasibility of an efficient validation to check its fulfilment.

3.2.2 Quality factors for DSM

As recalled above, the quality of a digital surface model is affected by both point location accuracy and resampling efficiency.

Point location accuracy

Each point location technique has a specific error budget, with contributions from the intrinsic features of the employed system (sensor, platform, etc.) and from the acquisition and processing parameters.

Most DSM extraction techniques (and particularly photogrammetric and profiling techniques) are direct techniques, which means that they provide point location measurements in which adjacent samples are independent from each other. They are more suitable for elevation mapping than for slope mapping. On the contrary, differential techniques (such as shape from shading or radar interferometry) provide information about surface orientation, i.e. slope or azimuth. In this case, elevations are obtained by slope integration, so that height errors are subject to propagation. This phenomenon may be reduced by crossing different integration paths.

As far as resampling is concerned, its impact on DSM quality depends on grid geometry (structure and density) and on data format.

Grid structure

A great variety of grid structures has been proposed for digital surface model resampling. They have been reviewed and discussed by several authors (Burrough, 1986; Carter, 1988). Three main approaches can be mentioned:

- regular sampling, in which all meshes have constant size and shape (most often square);
- semi-regular sampling, which is based on a very dense regular grid in which only useful points have been selected;
- irregular sampling, where terrain points may be located anywhere.

Regular sampling has obvious advantages in terms of storage, but it is not very efficient to depict natural relief in which shapes and textures are mainly irregular, unless the grid is considerably densified. This is the reason why semi-regular methods like progressive or composite sampling (Burrough, 1986; Charif and Makarovic, 1989), and irregular ones like TINs (Burrough, 1986; Chen and Guevara, 1987) are so successful for digital terrain modelling.

Grid density

The density of a DSM is basically a trade-off between economical constraints (which in general tend to limit the density) and accuracy requirements (which are rather fulfilled by a higher grid density).

The impact of mesh size on DSM quality has mainly been studied in terms of height error (e.g. by Li, 1992), but its major effect is on height derivatives. Indeed, reducing the density of a DSM grid (i.e. subsampling) removes the steepest slopes and makes the surface model smoother.

Data format

The digital format of the data has to be mentioned as well within the contributors to DSM quality, and in particular the number of bytes per sample. Formats using 2 bytes (i.e. 65,536 levels) which allow a 10 cm precision over an elevation range of more than 6,000 m, are commonly used because they provide a good trade-off between 1 byte (which limits the accuracy) and 4 bytes (which increases the volume of data needlessly).

3.2.3 Internal validation

A set of 3D coordinates drawn at random has very little chance of yielding a realistic relief. Therefore, it is worth checking that the terrain described by the DSM is possible, i.e. that it fulfils the main properties of real topographic surfaces. These properties can be quite straightforward. For instance 'rivers go down' or, in urban areas, 'building walls are vertical'. Checking to what extent these properties are respected does not require reference data but only a generic knowledge of landscape features. For this reason, it can be called internal validation.

Visual artefact detection, which should always be done before DSM delivery, is the first level of internal validation.

Unrealistic textures, such as strips or other anisotropic features, can be revealed by Fourier analysis or by comparing variograms in different directions. These artefacts can result from image scanning (Brown and Bara, 1994) or from contour line interpolation (Polidori *et al.*, 1991).

If some rules are universal (e.g. rivers go down), others require an expert knowledge about geomorphology (for natural relief mapping) or about urban structure design (for building extraction). For instance, texture isotropy is not supposed to be fulfilled everywhere, so that some subjectivity is required to distinguish natural anisotropies from directional artefacts. This distinction will be easier if the DSM extraction algorithms are known, because each extraction step may have a contribution to the observed artefacts: correlation noise, original image striping, grid resampling, etc.

3.2.4 External validation

If external elevation data are available and reliable, an external validation can be considered, which consists in comparing the DSM with the reference data. This is the most usual way of evaluating the quality of DSMs, but it is limited by two major difficulties.

The first difficulty is the availability of a suitable reference data set. Indeed, DSMs are often validated with very few ground control points, so that the comparison may be statistically meaningless. Moreover, these control points have their own error which in most cases is not known and which may have the same order of magnitude as the DSM they are supposed to control.

The second difficulty is the need for an explicit comparison criterion, which must reflect application requirements. On the one hand, the magnitude to be compared has to be defined: it is altitude in most cases, but slope or other derivative magnitudes could also be considered. On the other hand, a statistical index has to be defined too, generally based on the histogram of height differences: mean, standard deviation, maximum error, etc. have different meanings.

An interesting way of overcoming these difficulties is to map the discrepancies in order to display their spatial behaviour. This validation approach is not quantitative, but it proceeds very useful information, such as which landscapes (in terms of relief shape but also land cover) are accurately depicted and which are poorly shown. Therefore, it may contribute to improve the understanding of a given DSM extraction technique.

References

Brown D., Bara T. (1994) Recognition and reduction of systematic error in elevation and derivative surfaces from 7 1/2 minute DEMs. *Photogrammetric Engineering and Remote Sensing*, vol. 60, no. 2, pp. 189–194.

Burrough P. (1986) *Principal of geographic information systems for land resources assessment*. Oxford University Press, New York.

Carter J. (1988) Digital representations of topographic surfaces. *Photogrammetric Engineering and Remote Sensing*, vol. 54, no. 11, pp. 1577–1580.

Charif M., Makarovic B. (1989) Optimizing progressive and composite sampling for DTMs. *ITC Journal*, vol. 2, pp. 104–111.

Chen C., Guevara J.A. (1987) Systematic selection of very important points (VIP) from digital elevation model for constructing triangular irregular networks. *Proceedings of Auto Carto 8*, Baltimore, 29 March–3 April, pp. 50–56.

Fisher P. (1991) First experiments in viewshed uncertainty: the accuracy of the viewshed area. *Photogrammetric Engineering and Remote Sensing*, vol. 57, no. 10, pp. 1321–1327.

Lee J., Snider P., Fisher P. (1992) Modelling the effect of data error on feature extraction from digital elevation models. *Photogrammetric Engineering and Remote Sensing*, vol. 58, no. 10, pp. 1461–1467.

Li Z. (1992) Variation of the accuracy of digital terrain models with sampling interval. *Photogrammetric Record*, vol. 14(79), pp. 113–128.

Polidori L., Chorowicz J., Guillande R. (1991) Description of terrain as a fractal surface and application to digital elevation model quality assessment. *Photogrammetric Engineering and Remote Sensing*, vol. 57, no. 10, pp. 1329–1332.

Polidori L., Chorowicz J. (1993) Comparison of bilinear and Brownian interpolation for digital elevation models. *ISPRS Journal of Photogrammetry and Remote Sensing*, vol. 48(2), pp. 18–23.

3.3 3D DATA ACQUISITION FROM VISIBLE IMAGES

Nicolas Paparoditis, Olivier Dissard

Introduction

The surface that can be processed from images is an observable surface. In the case of aerial visible images, this surface will describe the terrain but also all the structures lying on the terrain at the scale/resolution of the images e.g. buildings, vegetation canopy, bridges, etc.

The techniques presented here are applicable to visible images of all platforms (helicopters, airplanes, satellites, ground vehicle, ground static, etc.) and of all kinds of passive sensors (photos, CCD pushbroom, CCD frame cameras) as long as the internal, relative and external parameters for each view are known. In other words, we suppose that the platform-triangulation process has already been achieved previously. Nevertheless, even though the concepts remain the same, the 3D processes and the integration of the processes themselves can change slightly, depending on the survey acquisition possibilities and constraints. We will point out whenever the particularities of the sensor and the platform have an impact on the processes described.

Besides the particular process assessment induced by the particular geometry and distribution of the data, the methods and the strategies to be involved also depend on what kind of 3D data we want to derive from the visible images, i.e. single-point measurements or more dense and more regular for the generation of a DEM. We will start from the simplest, the stereopair processing of an individual 3D measurement, before looking at the global processing for the generation of an elementary stereopair DEM and a global DEM over the survey, and we will end by defining the proper specifications for a survey acquisition system (and the processes to manage them) enabling the processes to avoid in a clean way all (or at least most) of the problems and allowing us to reach the automation of a complete and robust enough DEM on the whole survey.

3.3.1 Formulation of an automatic individual point matching process

Nicolas Paparoditis

The major interest of digital photogrammetry, besides all the management facilities that it provides, is that images are described by a set of regularly spaced digital values that can be easily manipulated with algorithms. Let I_1 and I_2 be the two digital images composing the stereopair, (i, j) a point position in one of the image grids, and $R(i, j)$ the corresponding grey-level value.

The concern of digital matching, is how do we determine that any point (i_1, j_1) of I_1 and point (i_2, j_2) of I_2 are alike or homologous? And more generally, how do we determine that characteristic features of I_1 and I_2 are alike? This preoccupation is one of the most recurrent in digital photogrammetry, image processing and in automatic pattern recognition techniques.

Entity attributes and similarity measure

As we are looking for individual point matches, we will stick for now with point entities. Most of the time a function f derives from image measurements (values) on or around our point entities (i_1, j_1) and (i_2, j_2) corresponding entity attribute vectors $\mathbf{V}_1$ and $\mathbf{V}_2$. Then a similarity function g (g can be a distance but does not have to be) derives a numerical similarity value (or score) C describing the similarity between $\mathbf{V}_1$ and $\mathbf{V}_2$ and thus the entities themselves. C can be expressed as follows:

$$C((i_1, j_1), (i_2, j_2)) = g(\mathbf{V}_1, \mathbf{V}_2).$$

Interpolation function

We point out that the image sampling effects are such that the real homologue (i_2, j_2) of a point (i_1, j_1) on the grid of image 1 is not itself on the grid in image 2. Thus the function f should implicitly use and be built on an interpolation function $\mathbf{V}$ (bilinear, bicubic, sinc, etc.) which determines the intensity value in any sub-pixel point of an image from its closest neighbours on the image grid.

An ill-posed and a combinatorial problem

In practice, how do we build **V** and choose f? Let us take as an example the simplest f and g. For instance, f giving the intensity value itself and g the difference between the intensity values. Due to the image noise, the radiometric quantification, and the image sampling, the grey-level values themselves for real homologous points are in general different. Thus, many points in I_1 will have close intensity values and for a given point in I_1 many points in I_2 will have close intensity values. Thus we have a serious matching ambiguity problem and we can conclude that the information derived from f is not characteristic enough and g not stable and discriminating enough. We also face a combinatorial explosion especially if we want to match all points of I_1.

The problem of digital image matching belongs to the family of mathematically ill-posed problems (Bertero et al., 1988). The existence, the uniqueness, the stability of a solution are not a priori guaranteed. This problem can be transformed in a well-posed one by imposing regularizing constraints to diminish the degrees of freedom in the parameter space so that the domain of possible solutions can be reduced (Tikhonov, 1963).

Area-based matching: a first solution to ambiguity

One solution to this matching ambiguity problem is to make the entities attribute vectors as unique as possible. Instead of matching the grey level of the pixel, we can match the context (all the grey levels) around the homologous points. We here make the assumption that the contexts are also homologous. The context is most of the time characterized by all the grey levels of the pixels inside a rectangular (most of the time square) template window (also called image patch) centred on the entity thus with

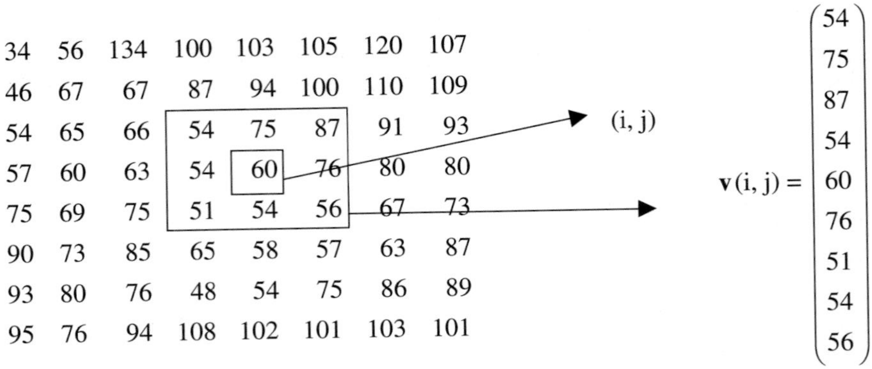

$$\mathbf{v}(i,j) = \begin{pmatrix} 54 \\ 75 \\ 87 \\ 54 \\ 60 \\ 76 \\ 51 \\ 54 \\ 56 \end{pmatrix}$$

Figure 3.3.1

an odd number of lines N and rows M. The entity can be described by a vector v with $N \times M$ components. Each component is a grey level of a template pixel following a given spatial ordering (as shown in Figure 3.3.1). Thus, the vector describes the local image texture pattern, which is of course much more discriminating than a single intensity value. The larger the templates, the higher the discrimination power. Meanwhile, our context stability assumption becomes less and less valid (see 'Geometric assumptions' hereunder).

Matching the entities is equivalent to matching their area context vectors. Matching these vectors is what we call area-based matching. The requirement for area similarity functions is that their response is optimal for real homologous points and that they are as stable as possible to some radiometric and geometric changes between homologous template areas in both images. The most commonly used similarity functions on these areas are the following:

Least squares differences:

$$C_1((i_1, j_1), (i_2, j_2)) = \left\| \frac{V_2(i_2, j_2)}{\|V_2(i_2, j_2)\|} - \frac{V_1(i_1, j_1)}{\|V_1(i_1, j_1)\|} \right\|^2 ; \qquad (3.4)$$

Scalar product:

$$C_2((i_1, j_1), (i_2, j_2)) = \left\| \frac{V_1(i_1, j_1) V_2(i_2, j_2)}{\|V_1(i_1, j_1)\| \, \|V_2(i_2, j_2)\|} \right\|^2 . \qquad (3.5)$$

As shown in Figure 3.3.2, these similarity functions have a geometrical explanation. C_1 is the norm of V_3 and C_2 is $\cos \theta$, where θ is the angle between V_1 and V_2. When the textures are alike V_1 and V_2 should be nearly collinear. Thus C_1 should be close to 0 and C_2 close to 1.

If corresponding templates have the same texture but are affected by a radiometric shift (case where $V_1 \sim V_2 + \alpha$), due to the fact that the images could be acquired with different lighting conditions or in the case of

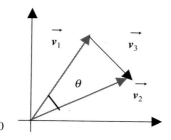

Figure 3.3.2

analogue photographs that the images have been scanned in different conditions or that the observed surface has a non-Lambertian behaviour, the similarity scores will decrease significantly thus altering the discrimination process. To overcome this problem the vectors can be centred. Thus if μ_1 (resp. μ_2) is the mean grey level and σ_1 (resp. σ_2) is the root mean square of grey levels of the template around (i_1, j_1) (resp. (i_2, j_2)) the previous similarity functions become:

Centred least squares:

$$C'_1((i_1, j_1), (i_2, j_2)) = \left\| \frac{V_2(i_2, j_2) - \mu_1}{\sigma_1} - \frac{V_1(i_1, j_1) - \mu_2}{\sigma_2} \right\|^2 ; \qquad (3.6)$$

Linear cross-correlation:

$$C'_2((i_1, j_1), (i_2, j_2)) = \frac{V_1(i_1, j_1) V_2(i_2, j_2) - \mu_1 \mu_2}{\sigma_1 \sigma_2} . \qquad (3.7)$$

Could not the images be pre-processed with a global histogram radiometric equalization instead of centring each vector? Indeed not! This process would alter the image quality and consequently the quality of the matching process. Furthermore, some of these effects change through the images, i.e. hot spot and non-Lambertian effects. Could the contrasts be enhanced? No! In general we do so when we look at an image, because of our non-linear and low differential eye sensibility. Nevertheless, a correlation process has no such problems.

Image matching definition

Under the assumption the matching problem is well posed, if (i_2, j_2) is the homologue of (i_1, j_1) inside the admissible domain of hypotheses S within I_2 then one of the two following expressions should be verified for a given similarity function C:

$$(i_2, j_2) = \underset{(i,j) \in I_2/S}{\mathrm{ArgMax}}\ C\big(V_1(i_1, j_1), V_2(i, j)\big)$$

or

$$(i_2, j_2) = \underset{(i,j) \in I_2/S}{\mathrm{ArgMin}}\ C\big(V_1(i_1, j_1), V_2(i, j)\big) . \qquad (3.8)$$

Geometric assumptions

The area-based matching process described above rests upon a major assumption. Indeed, we have implicitly supposed that the homologous template of image 1 can be found in image 2 by a simple shift. In general, the differences in the viewing geometry (due to erratic variations of attitude of the sensor) between the two views and the deformations due to the landscape relief are such that this assumption is not strictly verified.

Feature-based matching

If the point entity itself is or belongs to a characteristic image feature, e.g. an interest point (see §2.4) or a contour point, the homologous point can be looked for in the set of identical image feature entities in the other image. This will restrict the combinatorial problem and thus the ambiguity problem. Furthermore, these entities are less sensitive to geometric deformations between the images. Information such as local gradient norm or direction, second or higher degree image derivatives can be used to build the feature entity attribute vectors. Nevertheless, in an urban landscape for instance, many image features look alike and finding characteristic and discriminating entity attributes is difficult and sometimes impossible. Moreover, the entities themselves, e.g. building contours, are not stable thus the ambiguity problem remains. Feature-based matching is also an ill-posed problem and is limited to a sparse set of points in the scene and thus does not provide the universality of area-based techniques for matching every point in the scene or in the images.

Stereopair geometric constraints: a second solution to the ambiguity problem

Another way of limiting the ambiguity problem is by reducing the combinatory of the matching problem, i.e. the search space for homologous points. Indeed, the bigger the search space the higher the number of possible homologous hypotheses and the higher the probability of encountering matching ambiguities.

The following paragraphs describes the different matching methods that can be applied to an oriented stereopair to overcome the geometric distortion problems and to restrict the spatial domain S of homologous solutions. Generally the point-to-point matching problem in a stereopair geometry can be expressed in two radically different ways and thus leads to two different matching methods each having different advantages.

3.3.2 Stereo matching from image space (SMI)

Nicolas Paparoditis

This first method only requires the relative orientation of the images for the matching process. If the real 3D localization is meant to be computed, the absolute orientation of the images are required. The matching

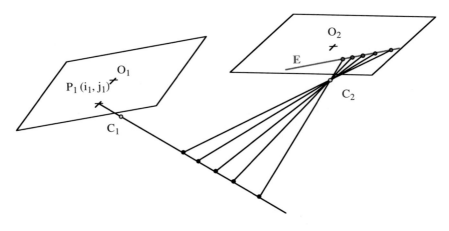

Figure 3.3.3

formulation of this method, which is image-based, can be expressed in the following terms: 'given a (i_1, j_1) in I_1 (called the master or reference image) what is the corresponding (i_2, j_2) in I_2 (called the slave or secondary image)?'

Epipolar lines

The position of (i_2, j_2) is geometrically (stereoscopically) constrained by (i_1, j_1). As we can see in Figure 3.3.3, all possible matches are on a line denoted E in the image 2, called the epipolar line, which is the projection in the image plane of the 3D plane lying on the 3D ray of (i_1, j_1) including C_1 and on C_2. Due to geometrical distortions of the optics, this line is more likely to be a curve. Nevertheless, the general 2D matching problem is now transformed into an easier 1D matching problem where (i_2, j_2) has

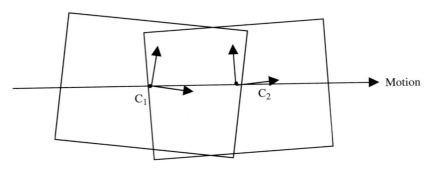

Figure 3.3.4

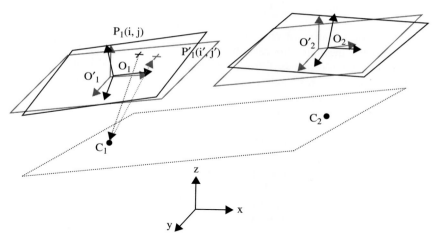

Figure 3.3.5

to be searched for along this curve. Note that all epipolar lines corresponding to all points of image 1 intersect themselves in a point of image 2, called epipolar point, which is the projection of C_1 in the image 2 plane (see Figure 3.3.4).

If I_1 and I_2 are two consecutive images in the same aerial flight strip (near to vertical viewing axis), or classical across track (up to 50° of latitude so that we can consider the orbits to be locally parallel), or along track satellite stereopairs, homologous texture vectors of a given landscape patch (with a low slope) can be found by a simple translation. Thus our previous correlation scores similarity criteria can be applied directly. In these cases, the epipolar lines are very close to the image lines themselves.

Nevertheless, for some aerial surveys, the differential yaw – but also pitch and roll – angles between both views, due to plane instability, can sometimes occur. The higher the difference is and the larger the window size gets, the less the texture vectors for homologous points will look alike, and the more the automatic matching process is unlikely to give good results. In these conditions to overcome this problem, a gyro-stabilized platform can be used so that epipolar lines follow as closely as possible the image lines (see Figure 3.3.5).

Epipolar resampling

To ease the algorithmic implementation of the matching process, the images are often resampled in a way that the epipolar lines become parallel and aligned on the image lines. Let I'_1 and I'_2 be the resampled images. Thus looking for the homologous estimate of point (i_1, j_1) can now be expressed:

$$(\hat{i}_2, \hat{j}_2) = (\hat{i}_2, j_1) = \underset{x \in [0, N]}{\text{ArgOpt}}\, C\left(V_{I_1'}(i_1, j_1), V_{I_2'}(x, j_1)\right), \tag{3.9}$$

where N is the number of rows of I_2'.

How can we manage this resampling? Let (O_1, i_1, j_1, k_1) and (O_2, i_2, j_2, k_2) be the two reference systems describing the orientation of the image planes. We simulate for each view the image of a virtual sensor (I_1' and I_2') with the same optical centres but with new reference systems and without distortions of the optics and of the focal plane. The new reference systems (O_1, i_1', j_1', k_1') and (O_2, i_2', j_2', k_2') are constructed so that i_1' and i_2' are collinear to C_1C_2, k_1' and k_2' are collinear to z, and $j_1' = k_1' \wedge i_1'$ and $j_2' = k_2' \wedge i_2'$.

How do we construct the pixel grey levels of these new images? For each pixel $P'(i', j')$ of image I', we determine with our new sensor's geometry the corresponding ray in 3D taking into account the distortion of the camera. According to previous geometry, we determine the corresponding image position $P(i, j)$ in image I by intersecting the 3D ray with its image plane. Of course, this point falls anywhere inside the image grid, thus the grey-level value for $P'(i', j')$ has to be calculated by interpolation of the grey-level values of neighbour nodes of $P(i, j)$ inside the image grid.

If it can be avoided, in the case of classical good conditions surveys, it is preferable. Indeed, this operation leads to a degradation of image quality which will alter slightly the quality of the image matching process. The only advantage of a systematic epipolar resampling is that the images can be viewed directly in digital photogrammetric workstations whatever the differential yaw between the images. (See Figure 3.3.6.)

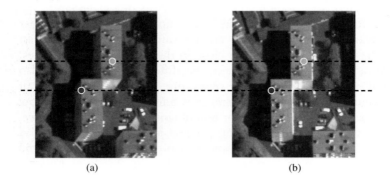

(a)   (b)

Figure 3.3.6 Epipolar matching. The left image (a) and the right image (b) extracts are the result of the stereopair resampling. The broken lines correspond to the horizontal conjuguate epipolar lines in these resampled images.

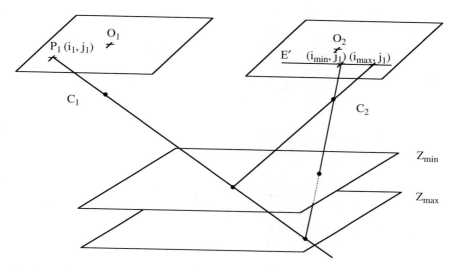

Figure 3.3.7 Defining a search space interval.

Defining an interval for the match search

The range of positions for all possible homologous hypotheses for a given $P_1(i_1, j_1)$ in I_1 can be delimited inside an interval $([i_{2min}, i_{2max}], j_1)$ on the epipolar E' by the knowledge of minimal and maximal z values (or the range of disparities/parallax if we only have the relative orientations) on the scene (as shown on Figure 3.3.7) if we want to have a given disparity range for all the points in image I_1, or a gross relief model if we want to have this information more locally. This information could be provided by existing databases, e.g. world-covering low-resolution DTM or by a map.

Another way of doing this, is to stereoplot manually the homologous image points for the two landscape points corresponding to the lowest and the highest point in the scene or plotting a larger set of points regularly sampling the scene if we want to have a locally adaptive range definition. Some more elaborate ways of reducing these intervals will be given in §3.3.8.

Search-space sampling and homologous position estimation

If d_i is the spatial sampling in I_2 along the search space interval $[i_{2min}, i_{2max}]$, all the correlation scores for all the possible moves form a graph usually called a correlation profile. The homologous point is considered to be the point where the maximum of the similarity score (also called the correlation peak) occurs (see Figure 3.3.8). Its position can be expressed as follows:

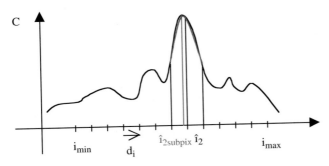

Figure 3.3.8 Sub-pixel localization of the correlation peak.

$$\hat{i}_2 = \underset{k \in [0,(i_{max}-i_{min})/di]}{\mathrm{ArgMax}} C_2\big(V'_1(i_1, j_1), V'_2(i_{min}+k \cdot di, j_1)+i_{min}\big) \, . \quad (3.10)$$

The smaller the d_i the better the correlation profile is sampled and the higher the matching precision. In practice, to limit the number of similarity scores calculation, most of the time d_i is fixed to 1 so that all the homologous points hypotheses fall on the grid and thus the texture vectors are directly extractable from the images without any further interpolation processing. If we consider the peak to be the integer position of the homologous point, the maximum error in the homologous estimation is 0.5 pixels. A more precise (in most cases) sub-pixel position can be found by fitting a polynomial function, often a parabola, through the correlation values around the peak position. The most commonly used fitting function is the parabola. With the parabola, the sub-pixel position of the correlation peak can be obtained from the discrete positions as follows:

$$\hat{i}_{2\text{Subpix}} = \hat{i}_2 - \frac{C(\hat{i}_2+1, j_1) - C(\hat{i}_2-1, j_1)}{2\big(C(\hat{i}_2-1, j_1) + C(\hat{i}_2+1, j_1) - 2C(\hat{i}_2, j_1)\big)} \, . \quad (3.11)$$

If the d_i sampling is sub-pixel, with the idea of finding directly a correlation peak closer to the real homologous position without any *a posteriori* interpolation, then the homologous hypotheses samples fall inside the grid and thus all the components of the texture vectors have to be interpolated. Generally a d_i of one-fifth of a pixel reaches the limit of the matching process precision. One should remark that the grey levels of texture vectors of image I_2 will have gone through two interpolation processes. We have pointed out earlier that a resampling alters the image quality and consequently the matching quality. In this particular case, this direct sub-pixel matching process should be carried out on the non-resampled images along the epipolar curve E.

Doing the interpolation on the images or on the correlation profile, provides very close results. If the matching process has to be carried out on many points of image I_1, one would prefer using the first technique as the number of samples to be calculated is smaller.

Managing with relative orientation errors

Due to the relative orientation errors, the real homologue of (i_1, j_1) is not on the computed epipolar line E' itself but in a close range of the epipolar line. To cope with that the search space can be extended to all points of image I_2 in a neighbourhood (depending on relative orientation residues; 1 pixel is usually enough) on each side of the epipolar line. Thus our correlation profile becomes a correlation surface. If the peak of the correlation surface occurs for a point on the epipolar line, the easiest way of finding the sub-pixel position is to first determine the sub-pixel position in the epipolar direction by fitting a parabola through the correlation scores and then finding the sub-pixel position in the orthogonal direction by fitting another parabola. If not the neighbourhood should first be extended before a further process.

3D localization of homologous points

The 3D localization of homologous points (i_1, j_1) and (i_2, j_2) is obtained by intersecting both corresponding 3D rays $D_1(C_1, \mathbf{R}_1)$ and $D_2(C_2, \mathbf{R}_2)$. If (i_2, j_2) does not lie exactly on the epipolar line of (i_1, j_1), those rays do not intersect, as shown on Figure 3.3.9.

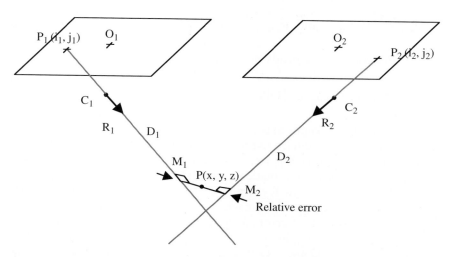

Figure 3.3.9 3D triangulation and localization of homologous points.

Let M_1 be the closest point of ray D_1 to ray D_2 and M_2 the closest point of ray D_2 to ray D_1. M_1 and M_2 are such that:

$$\begin{cases} M_1 = C_1 + \lambda_1 R_1 \\ M_2 = C_2 + \lambda_2 R_2 \\ M_1 M_2 \cdot R_1 = 0 \\ M_1 M_2 \cdot R_2 = 0 \end{cases} \text{ thus } \begin{cases} \lambda_1 = \dfrac{(C_1 C_2, R_2, R_1 \wedge R_2)}{(R_1 \wedge R_2)^2} \\ \lambda_2 = \dfrac{(C_1 C_2, R_1, R_1 \wedge R_2)}{(R_1 \wedge R_2)^2} . \end{cases} \quad (3.12)$$

The corresponding 3D point is generally chosen to be $P = (M_1 + M_2)/2$, the closest and equidistant point to both rays. The $\|M_1 M_2\|$ distance gives us a quality estimator of the 3D relative localization.

Relation between 3D precision and matching precision

The 3D localization precision is affected by the aerial triangulation errors, the matching errors and its planimetric and altimetric components depend on the viewing angles and on the stereoscopic base to height ratio (B/H) as shown on Figure 3.3.10. Putting aside the errors due to the aerial triangulation process, thus making the assumption that the relative orientation is perfect, we can estimate theoretically the intrinsic precision of the matching process. Let e_{match} (resp. e_{alti}) be the matching (resp. altimetric) error and σ_{match} (resp. σ_{alti}) the root mean square of the matching (resp. altimetric) error expressed in pixels and r_0 the ground pixel size in metres. Thanks to the Thales theorem, the altimetric error is given by:

$$e_{alti} = \frac{H}{B} r_0 e_{match} \text{ and } \sigma_{alti} = \frac{H}{B} r_0 \sigma_{match}.$$

The planimetric error depends on the position of (i_1, j_1) in image I_1:

$$e_{plani} = \tan(i) \cdot e_{alti} = \|OP_1\| \frac{e_{alti}}{f},$$

where f is the focal length expressed in pixels.

The formulas that we have expressed here are not the models of the 3D errors that we would have if we intended plotting a given ground feature in image I_1 and finding automatically the 3D corresponding point. Indeed the matched point (i_2, j_2) will correspond to (i_1, j_1) and not to the feature that was meant to be plotted in image I_1. This plotting error will inevitably lead to another source of error in the 3D localization. Only the plotting errors along the epipolar lines will influence the altimetric precision. Let the parallax be $\mathbf{p} = O_2 P_2 - O_1 P_1$ and its errors along the epipolar lines $d\mathbf{p} = d\mathbf{p}.\mathbf{i} = e_{match} - e_{plotting} \mathbf{i}$ then the errors are given by:

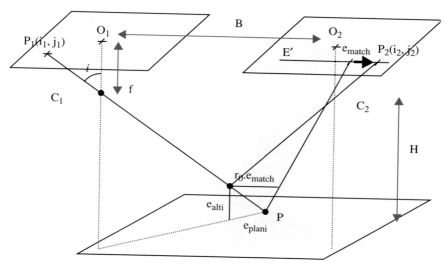

Figure 3.3.10

$$e_{alti} = \frac{H}{B} r_0 \, dp \quad \text{and} \quad \sigma_{alti} = \frac{H}{B} r_0 \sqrt{\sigma_{match}^2 + \sigma_{plotting}^2} \, . \quad (3.13)$$

The higher the base to height ratio, the better the altimetric precision. The choice of this ratio for a survey is conditioned by the desired quality for the mapping output. For a DEM production line, for instance, a higher one should preferably be chosen. Against that, when the ratio increases, the image differences and consequently the matching difficulty increases. (See Figure 3.3.11.)

Intrinsic matching precision

What is the precision that we can achieve from digital automatic matching? The limits of the matching precision are easy to estimate in the best case, i.e. when the homologous pattern can be found by a simple translation (horizontal images planes, flat landscape). To generate a virtual global translation between I_1 and I_2, we can simulate a horizontal image plane stereopair with perfectly known orientation parameters on a perfectly flat landscape where we can map any kind of image pattern, e.g. an aerial orthophoto. Since we have simulated the images, we know exactly the translation value between I_1 and I_2. If we apply our matching process to estimate these translations on different stereopairs simulated for different translation values (fractional part between 0 and 1), we find systematic errors in the translation estimation depending on the real translation fractional part as shown in Figure 3.3.12.

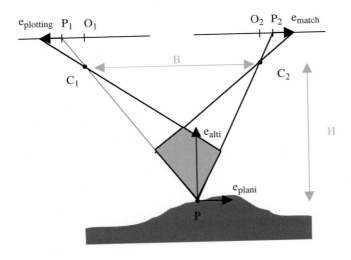

Figure 3.3.11

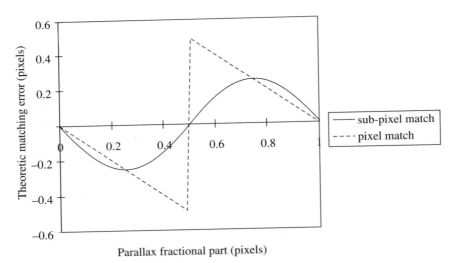

Figure 3.3.12 Intrinsic matching errors.

This bias is due to the interpolation process and to the interpolation function. We still have a bias if we do not use a sub-pixel estimator. Indeed, if we do so we still implicitly use an interpolation function which is the nearest-neighbour interpolator.

With the sub-pixel estimation process, the maximum bias occurs for a sub-pixel translation of 0.25 pixels. The amplitude of this bias depends on the interpolation function. It can range from 0.15 to 0.25 pixels. This does not mean that the function giving the best theoretical precision is the best in practice. In general, the simpler the interpolation functions (using the smaller neighbourhoods) the more robust they are for real image digital matching. From now on, we will consider that the average matching precision is of a quarter of a pixel.

In the case of digital manual stereo-plotting, the matching precision is difficult to evaluate statistically because of all the factors to be taken into account: the radiometric (contrast), geometric (punctual, linear, corner, etc.) and 3D context characteristics of the feature to be plotted, the image quality (noise, blur, etc.), the stereoscopic acuity and the tiredness of the operator, the optical quality of the stereo display, the sub-pixel possibilities of the stereo display (sub-pixel image displacement, etc.), and the plotting methodology. But in fact, most of these difficulties are present with automatic matching techniques too.

3.3.3 Stereo matching from object space (SMO)

Nicolas Paparoditis

Photogrammetrists may express the matching problem for the 3D localization in a completely different way. In fact, the other way round: 'given a (X, Y) in object space what is the corresponding z?'. As we can see in Figure 3.3.13, for a given (X, Y) if we change the z value, the displacements of the corresponding image points are implicitly constrained along the curves which are the projection of the planes lying upon each nadir axis and (X, Y, z). Looking for the best z, is thus looking for the best image match along these 'nadir' curves. Thus as for the image matching process guided from image space, we build a correlation profile but instead of computing all correlation scores for all the possible parallaxes along the epipolar line we compute all correlation scores for all possible z values in a $[z_{min}, z_{max}]$ search space interval. The best elevation estimate is given by:

$$\hat{z}(X, Y) = \underset{z \in [z_{min}, z_{max}]}{\mathrm{ArgMax}}\; C_2\big(V_2(\mathrm{Loc}_{I_2}(X, Y, z)), V_1(\mathrm{Loc}_{I_1}(X, Y, z))\big), \quad (3.14)$$

where

$$\mathrm{Loc}_{I_1}: \Re^3 \to I_1 \;\; (x, y, z) \to (i_1, j_1)$$

and

$$\mathrm{Loc}_{I_2}: \Re^3 \to I_2 \;\; (x, y, z) \to (i_2, j_2)$$

are the object to image localization functions. One should remark that with this stereo method guided from object space, the image matching and

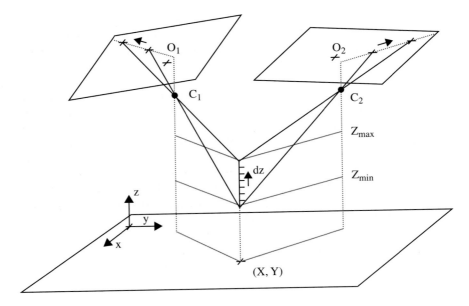

Figure 3.3.13

the 3D localization processes are performed at the same time. The elementary sampling dz along the vertical line can be determined to be:

$$dz = \frac{H}{4B} r_0, \qquad (3.15)$$

where $r_0 = H/f$ is the ground pixel size.

Resampling correlation templates

Due to the perspective deformations and to the relative rotation between image planes, the image patches corresponding to a flat square ground patch ds do not overlap. Thus to obtain the best and the most discriminating correlation score between those patches, the images are locally resampled.

Indeed, as shown in Figure 3.3.14, we build two ortho-templates centred on (X, Y). The image resolution of these templates is equivalent to that of the images. For each pixel of these two ortho-templates we determine the 3D position on the ground patch, and we assign to each of them the grey-level value for the corresponding image position. If our local flat ground assumption is correct these should geometrically be exactly overlapping.

Stereo matching from object space is the proper way of obtaining automatically 3D measurements in a stereo display. Indeed, the observation

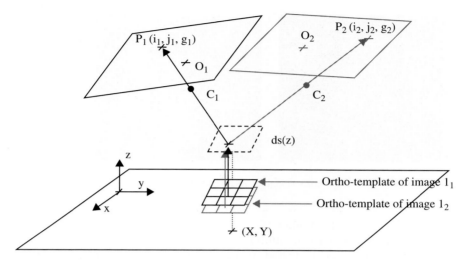

Figure 3.3.14

space in a stereo display is directly the object space. The operator can, in real time, obtain the elevation corresponding to a given plotter position (X, Y) or can plot a flat line and have the 3D profile along this line, e.g. for volume calculations or for road elevation profile estimation.

3.3.4 Difficulties due to image differences

Nicolas Paparoditis

Classical template matching suffers from a certain number of defaults, due to the non-respect of the assumptions it relies upon, that is to say the geometric and radiometric similarity of templates centred on homologous points and the discrimination power of correlation scores computed on the radiometric templates. Sometimes, the landscape properties and the geometric configurations of the image views are such that these assumptions are not respected, and thus the similarity, existence and uniqueness constraints for homologous points solution are violated:

- *On homogeneous areas and periodic structures.* The template radiometric content does not ensure a possible discrimination between all possible matches that fall in this area. The uniqueness constraint is violated. Having a higher image quality or a larger window definitely reduces the ambiguities for homogeneous areas. Periodic structures are a problem when the spatial periodicity orientation is along the

Figure 3.3.15 Homogeneous surfaces and repetitive structures.

baseline. Having another view not on the same stereo baseline will considerably reduce the ambiguities. (See Figure 3.3.15.)

- *On hidden areas.* In urban areas, for instance, all points observable in an image do not always exist in the other image. Especially in urban areas some zones are not seen on both images. Template matching is a low level process. Even if there are no possible matches for a given (i_1, j_1) or for a given (X, Y), we will still have a maximum peak in the correlation profile or surface which will lead to a false match. The existence constraint is violated. A way of bypassing this problem, is to have a higher stereo overlap (or more generally along or/and across track multi-stereo views) so that at least every point of the scene can be seen in at least two images, or to use multi-stereoscopic techniques. (See Figure 3.3.16.)
- *On non-overlapping areas.* The incomplete overlapping of the stereopair is one of the problems that occurs in digital stereo matching. Indeed, many points of both images having no homologous solutions will lead to problematic mismatches. The existence constraint is violated.

Figure 3.3.16 Mobile vehicles and hidden areas.

3D data acquisition from visible images 187

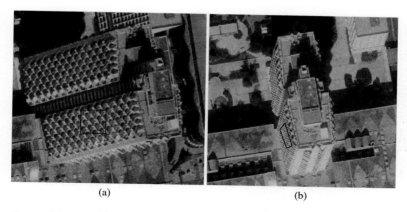

(a)   (b)

Figure 3.3.17 Differences in size and in shape of two real homologous patterns on steep sloped surfaces.

- *On steep sloped surfaces.* The image pattern shapes corresponding to a sloped planar ground patch in object space are geometrically not alike. These deformations are accentuated when the slope rises. Thus square image templates centred on homologous points have less probability of looking alike when their size increase. The similarity constraint is violated. A solution is having higher quality images so that we could take smaller windows, or coping with the deformations by using adaptive shape windows. (See Figure 3.3.17.)
- *Areas around 3D discontinuities.* When a template overlaps a 3D elevation, as all the pixels contribute to the matching process, the elevation determined in these areas will be a 'mean' of all 3D elevations inside the templates weighted by the contrast distribution. This will lead to a smoothing of the 3D relief morphology or to a spatial shift of the 3D structure of one half of the window size, as shown in Figure 3.3.18. We can overcome this problem using small window sizes thus assuming again high image quality, or using adaptive shape windows.
- *Non-Lambertian surfaces.* This happens for areas that have anisotropic surface reflection properties. If we observe this surface from different viewpoints, the radiometric measurements can be radically different. When the images are not saturated, and if the non-Lambertian effects are due relatively to a smooth surface compared to the ground pixel size, template matching can still be used if the texture vectors are centred to correct the radiometric bias between both homologous templates. When some areas are saturated (no details) in one of the images or when the non-Lambertian effects are due to 3D microstructures, for instance a cornfield seen under two different viewing angles, there is no satisfactory way of finding a good solution other

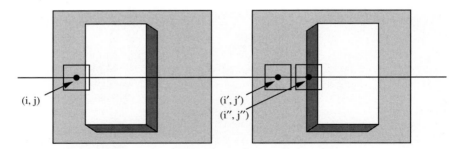

Figure 3.3.18 Matching areas around 3D changes. If we are looking for the homologous template of (i, j) along the epipolar line, the best correlation score will be obtained for (i'', j'') instead of (i', j'). Thus the parallax/elevation associated to (i, j) will be the one of the rooftop.

than rejecting the matches for these areas, unless we consider another stereopair if we have a survey with a high stereo overlap, or if we are in any multi-stereoscopic configuration.

- *Moving vehicles and shadows.* Matching problems can occur due to the fact that stereopairs are not instantaneous. Indeed, features like vehicles and shadows can move from one image to the other. Some vehicles appear or disappear, thus the existence constraint is violated. For some vehicles moving along the baseline, the matching process will generate aberrant corresponding 3D structures integrating the 3D and the movement parallax. For satellite or aerial across track stereopairs, shadows have moved thus the matching process can reconstruct 3D shadow limits at non-realistic elevations.

3.3.5 Coping with the problems: analysis of the correlation profile

Nicolas Paparoditis

One of the easiest solutions to coping with these problems is to be able to detect and reject all the matches where we think that the probability of encountering a problem is high. We can study the morphologic structure of the correlation profile and verify that it respects the properties of a good match, i.e. high score for the maximum peak, large correlation principal peak, no high secondary peaks. Many correlation profiles whatever the area do not respect this model. Indeed, some good matches have low and some others high correlation scores, and some false matches have high correlation scores. This is thus not an acceptable solution if we want this process to be universal. Of course, this kind of process can be applied as a last resort when no particular solutions can be found for each one of these area problems.

3.3.6 Coping locally with the problems: adaptive window shapes and sizes

Nicolas Paparoditis

The major problem of correlation template/area-based matching techniques is due to their shape rigidity (square or rectangular) and their fixed sizes. On the one hand, small templates enable an improved location of the good matches. On the other hand, larger windows allow a better discrimination between all possible matches. The problem is that we are confronted with a classical detection/localization compromise induced by the rigidity of the templates. Templates' sizes and shapes should thus change automatically and adapt themselves according to the changes in the landscape. We will now explain how this can be done.

Coping with homogeneous areas and periodic structures

For the SMI method and the homogeneous areas, the window size can be adapted to the local variance of the slave image contents having some knowledge about the image noise distribution. When areas are homogeneous, windows will be large, and when the areas are rich in texture, windows will be small. For periodic structures, we can choose the size of the template so that the auto-correlation of the texture in its own neighbourhood along the base line is mono-modal. This would mean that the texture comprised in the template is no longer a periodic pattern.

These adaptive size techniques are not possible with the SMO method. Indeed, we would correlate templates of different sizes and this would lead to heterogeneous correlation scores for all the elevations, thus choosing the correlation peak could lead to a false match. And it is difficult weighting the correlation scores according to template sizes. Nevertheless, one can verify that the templates corresponding to the correlation peak are textured enough to start, if necessary, another matching process on larger templates.

Another solution is to take a large window of fixed size for all the image and to weigh the grey-level values by a Gaussian distribution centered on (i_1, j_1). The interest of this technique is giving more importance to the pixels closer to (i_1, j_1). Nevertheless, if the area is globally homogeneous, an image feature inside the template window, even if it is relatively far away from (i_1, j_1), will allow a possible discrimination in the matching process. Besides, the estimated disparity will be the one corresponding to the feature and not to the central point of the template window.

Coping with hidden parts

For the SMI method, using the reciprocity constraint is an easy and efficient way of filtering false matches. We verify that the match of (i_2, j_2), homologous of (i_1, j_1), is also (i_1, j_1). This process is often called cross-validation. This filtering is very discriminating for false matches due to hidden areas. Indeed, if (i_1, j_1) belongs to a hidden area, it is most probable that the homologous of (i_2, j_2), estimated as the homologous of (i_1, j_1), will not be (i_1, j_1).

For the SMO method, we can do a similar process. For the elevation and the (i_1, j_1) and (i_2, j_2) corresponding to the correlation peak, we can verify with the SMI method that (i_2, j_2) is the homologous of (i_1, j_1) and vice versa.

Coping with discontinuities

We can adapt the templates locally if we have a priori information of where those 3D structures occur. If we do not have any maps, the only way of obtaining this information is by computing contour maps derived from the images assuming that a 3D discontinuity is often characterized in the images by a radiometric contrast. We can locally adapt the template shapes to the contours so they do not integrate patterns on the other side of the contours which could be areas at different elevation (Paparoditis et al., 1998). Cutting the template to produce a mask will of course reduce the textural content of the template, thus this process can only be carried out with large enough templates. These kinds of processes are half-way between area-based and feature-based matching techniques.

With the SMI method, for (i_1, j_1) we limit the template to all the pixels that are connected to (i_1, j_1), i.e. there is a non-contour crossing path joining, as shown in Figure 3.3.19(a). If a contour crosses continuously the template there will be no possible path joining the pixels on the other side of the contour. If a part of a structure corresponding to a 3D discontinuity is missing in the contour detection, the adaptive template will be the classical template itself. A solution is to constrain the path to be straight lines, as shown in Figure 3.3.19(b). The counterpart of this solution is that small open contours inside the template will mask all the pixels behind them and limit the number of pixels intervening in the calculation of the correlation score. A better solution to take into account those open contours is to weigh, as shown in Figure 3.3.19(c), all the grey-levels pixels given by the Figure 3.3.19(a) process by the inverse of the geodetic distance (Cord et al., 1998) which is given by the length of the shortest path joining a pixel inside the template and (i_1, j_1).

If (i_1, j_1) is itself on a contour, what should the template be: the right side or the left side of the contour? Neither in fact, the best solution in this case is feature-based matching, that is to say matching the contour

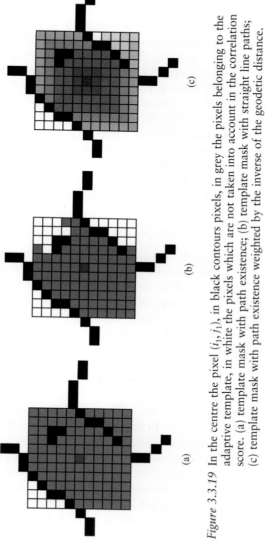

Figure 3.3.19 In the centre the pixel (i_1, j_1), in black contours pixels, in grey the pixels belonging to the adaptive template, in white the pixels which are not taken into account in the correlation score. (a) template mask with path existence; (b) template mask with straight line paths; (c) template mask with path existence weighted by the inverse of the geodetic distance.

itself. One solution is to look for all the contours in image I_2 along the epipolar line. Nevertheless, we need to remove ambiguity from all possible matches by considering the intrinsic information of the contour (we can verify that the contour orientations are the same) and the geometric and/or radiometric context around the contour (many solutions are possible). To add some robustness to this process, we can average, or take the median value of all disparities for all pixel edges locally along the contour line on each side of (i_1, j_1) assuming that the disparity is locally stable. These feature-based matching techniques are often not valid when the contours are not stable from one image to the other. This is the case for instance for building edges in urban areas, which are most often not characterized in the same way in both images.

With the SMO method, we consider the intersection of both image adaptive template shapes for a given elevation. The interest of this, compared to that of the SMI method, is that if a part of a 3D contour is missing in one of the two images, as long as the contours are complementary in both images this adaptive process will be efficient. Meanwhile this process is to be carried out carefully, as the number of pixels intervening in the correlation score calculation change all along the correlation profile.

Coping with geometrical distortions due to slopes

When image quality is good, small templates can be chosen to limit the distortion effects of sloped relief. In this case we may transform geometrically the homologous template (when guided from image space) or both templates (when guided from object space). We can model locally the relief by a polynomial function of a given order. Then the matching problem can be seen as finding the optimal set of coefficients giving the highest correlation score. When the order of the function increases, it is very difficult finding a robust solution and the process is much longer in time. So it is preferable to make stricter assumptions on the surface, i.e. supposing that the surface is locally planar.

Under these assumptions, for the SMI method we can determine for the square template around (i_1, j_1) in the reference image, the grey levels of the homologous template (same size and shape) by intersecting for each one of its pixels the corresponding 3D ray with a given patch hypothesis to determine the 3D corresponding point in object space. Given this 3D point we can determine the corresponding point in I_2 and interpolate the corresponding grey-level value we were looking for and then determine the corresponding image position and grey level (by interpolation) in the other image. For all the set of ground patches hypotheses, we can construct all the homologous templates and thus all the similarity scores. (See Figure 3.3.20, colour section.)

When guided from object space, we can calculate the correlation scores for all the ortho-templates built for each possible ground patch (z, α, β)

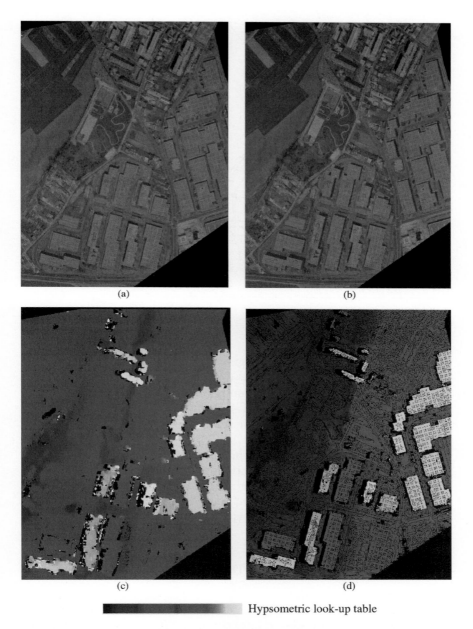

Figure 3.3.20 (a) Left image of an 1 m ground pixel size satellite across track stereopair; (b) right image; (c) DSM result with a classical cross-correlation technique; (d) DSM using adaptive shape windows.

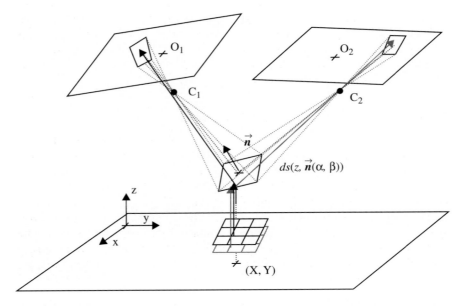

Figure 3.3.21

centred on (X, Y) as shown in Figure 3.3.21, α (resp. β) is the angle between the surface normal and x (resp. y). To limit the search-space possibilities we can make some assumptions on the maximum slope, e.g. the slope cannot reasonably exceed 45°. One should remark that implicitly the 3D entity we are looking for is not the 3D point on (X, Y) but an entire surface patch around (X, Y).

All these adaptive matching techniques can be mixed together. For instance, the slope and the 3D discontinuity techniques are to be mixed together if one wants to determine a 3D measurement of a point on a sloped roof next to the roof's edge. The combination of these two techniques is a strategy guided by the type of landscape to be mapped.

3.3.7 Strategies for a stereopair DSM generation

Nicolas Paparoditis

We discuss here the strategies of integration of different methods and techniques to build a regular grid DSM describing the observable surface. We do not propose an ideal and universal strategy of techniques integration but different ways of integrating the techniques and the problems related to their integration. Indeed, the strategy will depend on the type of landscape from natural smooth terrain to complex and dense urban areas, on the different image resolutions from low to very high, on the data acquisition conditions, and on image quality, etc. There is possibly a universal strategy, but it has not yet been found. Even if this strategy exists, some simpler strategies will be lighter in terms of computing times and efficient enough for a given set of data on a given landscape.

Basic matching entities and spatial sampling

Can we derive a DEM from a sparse set of 3D points corresponding to some matched image features entities, i.e. interest points, contour points, etc. in order to reduce the computing times? The underlying assumption would be that any relief point could be determined from the nearest 3D computed features by an interpolation process. Image features are sparse and do not sample the relief well enough to describe it correctly. Moreover, these entities are not stable and discriminating enough (especially in the case of urban scenes) thus leading to ill matches. The denser the entities, the finer the surface will be sampled. Points or surface patches are thus the basic matching entities for the DEM matching process even though image features can help in the process as we will see later on.

In the case of smooth natural relief, surface patches entities seem to be the most relevant to describe the surface. If we assume the relief to be smooth enough and planar locally with respect to the size of the patches, the DEM spatial sampling need not be that of ground pixel size. The DSM sampling should be constrained by our knowledge of the surface regularity. In consequence if the surface is smooth the ground size of the surface patches could be equivalent to the spatial sampling itself. Nevertheless, if we want the DSM to describe the surface in the most precise way, not only should the elevation be stored for every DSM grid point. Indeed, for many applications, the slope information is more or as relevant than the elevation one. So, the surface patch normal vector could also be stored for every grid point to improve the precision of both elevation and slope interpolation processes for a point inside the grid.

For 'rougher' surfaces, as in urban areas, a spaced grid of rectangular shape surface patches would not be adequate as the patches should be very small to describe the surface correctly. Thus the texture contents of the corresponding image templates will often not be rich enough to be entity discriminating. For urban surfaces, we prefer using a denser sampling, i.e. the process can be carried out for each pixel of image I_1 with the SMI method or for every grid point (X, Y) of the DSM with the SMO method. In the latter method, the DSM grid spatial sampling should be that of the size of the smallest objects we would like to represent. Meanwhile, this does not mean that the spatial resolution of the DSM is equivalent to the ground pixel size as the measured elevations are a 'mean' of all real elevations inside the template.

Initializing the matching process

A gross approximation or knowledge of the surface helps to reduce considerably the range of the search space intervals to diminish the computing times and to reduce the matching ambiguities probability. The initialization strategies are various.

Hierarchical matching

Hierarchical matching processes (Grimson, 1983) first aim at matching a very restricted subset of image entities i.e. characteristic image features such as, for example, interest points or contours and use these matches to initialize the matching of all the other image entities. The matching hypotheses for all the points lying between the matched features are searched for in a disparity/elevation search space interval framed by those of the matched features. This process is efficient when the feature entities are stable.

Using the tie points

A gross surface can be constructed using the tie points that have been used to determine the relative orientation of the images in the sensor viewing parameters determination process. The 3D samples corresponding to the matched tie points can be triangulated to construct a TIN surface that can be used to initialize the DSM generation. The image can be extremely dense and thus the gross surface can be quite close to the real surface.

Multi-resolution matching

Multi-resolution matching can be seen as a form of hierarchical matching. We first build for each image a multi-resolution pyramid as shown on Figure 2.5.3 using for instance a smoothing kernel filter, e.g. a Gaussian kernel or a wavelet decomposition. The matching starts from the lower level of the pyramid, (lowest resolution image) where the disparity vectors and the range of the search space are much smaller. The surface elaborated at this resolution is used to initialize the matching at the next step of the pyramid and so on. The major advantage of this strategy is to lower in a consequent way the computing times. This matching strategy is valid when the solution (DSM) provided for a given resolution is close to the surface observable at the highest resolution. It is thus valid for smooth surfaces but not for surfaces encountered in urban areas.

Object space vs. image space method

When the matching process is guided from image space, the 3D points corresponding to the matches of points located on a regular grid in slave image space have an irregular distribution in object space. Thus the generation of a spatial regular grid DEM goes through the blind resampling/interpolation of the irregularly distributed set of 3D samples, which inevitably leads to artefacts and alters the quality of elevations computed for all the DEM grid points. It is thus highly preferable determining the elevations of all the DEM grid points with the SMO method.

As a counterpart, some drawbacks in terms of practical implementation are to be mentioned. If we generate the DEM by computing the elevations sequentially for all the points some basic operations are highly redundant, thus the computing times will unnecessarily increase. Indeed, we mentioned earlier that the correlation score was applied for a given (X, Y) and for a given z to the ortho-templates generated by the local resampling of the images. Neighbouring points in the grid will have for the same z overlapping ortho-templates. Thus the resampling process for all pixels inside the overlap will be carried out in a redundant way. And the larger the windows the higher the redundancy. How can we avoid this? The solution is to resample both images in object space once for all for a given z. In other words we build two ortho-images (which have the same geometry as our DEM) assuming that surface is flat of elevation z. If the relief around a given (X, Y) is at an elevation z the ortho-templates extracted from these ortho-images are centred and (X, Y) should be alike. Thus for this given z and for all the (X, Y) of the DEM grid we can compute all the correlation scores.

In practice, the determination of the elevations for each DEM grid point can be done in two ways. The first one consists in initializing the DEM elevations to the lowest possible z and determining all the correlation scores for this z and for every point of the DEM grid with our global resampling and correlation process (we store the correlation scores in a corresponding buffer grid). For the next possible z we globally process the correlation scores in the same way. All DEM grid points where the new correlation scores are higher than those computed with the previous z have their elevation and correlation scores updated. We continue this process until we reach the highest possible z. If the storage capacities are limited, we can cut the global DEM into smaller ones and treat each of them independently and sequentially. The first interest is that we can verify that the correlation profile is a good-looking one. The second is that the correlation profile peak does not always identify the real solution which appears in the correlation profile as a secondary peak. Keeping all the information leaves us the possibility of choosing the best solution inside the peak set with a higher-level processing. This is what we are going to develop now.

Improving the process by looking for a global solution

We have up to now addressed the problem of independent 3D entity measurements. In the case of dense DEM measurements, we have generated a dense cloud of independent 3D entities that are supposed to describe a surface. These points should describe a 'possible' surface and neighbouring measurements should be coherent to a certain extent. Indeed, a 3D measurement cannot be completely decorrelated from its neighbouring measurements. Even when discontinuities occur their should be a correlation between all neighbours on each side of the break. We can thus verify locally or more globally the internal coherence of all the measurements assuming some modelled hypotheses or constraints on the surface basic properties.

Detecting erroneous 3D samples

A simple technique often applied to detect erroneous samples is based on the **ordering constraint also called visibility constraint**. This constraint, based on the assumption that the observable surface is a variety in the form of a single-valued function $z = f(x, y)$, verifies (for the SMI method) that three image points in a given order on an epipolar line also follow the same ordering on the conjugate epipolar line.

This constraint can also be applied, in object space, to the surface samples. Indeed, every 3D sample has been determined by matching homologous points thus supposing implicitly that these points are visible in both images. We can thus verify, with a z-buffering algorithm, that every 3D sample is directly visible in both images and not hidden by any other 3D samples. If some samples are hidden we have some evidence of the existence of erroneous samples. As shown in Figure 3.3.22 there is an obvious incompatibility between P_1 and P_2.

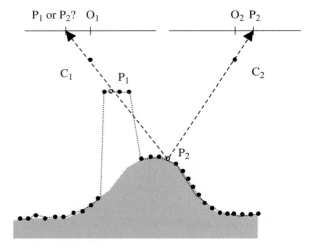

Figure 3.3.22 Mapped surface incoherence detection.

In the case of natural relief, we can verify that the surface is locally, smooth, planar and continuous. For a 3D patch process, we can verify that neighbouring 3D patches are connex. If one of the patches is not at all correlated with its neighbours, the patch is discarded and can be replaced by a patch fitting with the neighbouring ones. For a 3D point process, we can determine a plane fitting locally the 3D neighbour samples by a least squares estimation technique and verify that the considered point is close enough to this plane. We can still manage this process if a discontinuity occurs assuming that this discontinuity is characterized by a contour in one of the images. Indeed, we can transpose the 3D samples in image space and we can consider only the neighbouring samples that are on the same side of a contour. Besides, if a local connected set of 3D entities is erroneous, i.e. the errors are structured and spread, this process rejoins the ill-posed problems.

Finding the best 3D surface in a global way

A 'clean' solution to find the most 'probable' surface is keeping for all the entities all or a subset of the best (the highest correlation peaks) matching hypotheses and finding the best path or surface fitting the 'fuzzy' measurements by an optimization technique. We can evaluate the realism and the 'probability' of each possible path and retain the best one given a cost function integrating constraints on the surface behaviour. Figure 3.3.23

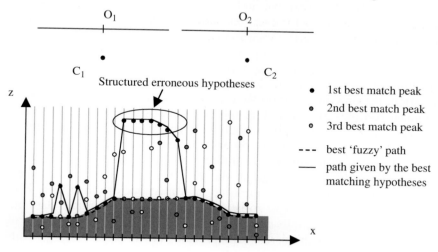

Figure 3.3.23 A virtual example of possible path optimization of the best 3D samples hypotheses given by the SMO method. In black we have the path given by the independent best matching hypotheses. In dotted line, we have the best 'fuzzy' path assuming some constraints, i.e. the surface is continuous and smooth with no steep slope.

shows an example of a global optimization with a continuity constraint where we have kept for each (X, Y) along a 3D ground profile, the three most probable z values corresponding to the three highest correlation peaks obtained with the SMO method.

The optimization techniques that can be applied are numerous, i.e. relaxation, least squares, dynamic programming, etc. We will not describe these techniques here, one can find them in any good signal or image processing book. What differentiates these techniques is the domain on which the optimization is carried out. Relaxation is usually applied to small neighbourhoods around the samples. Dynamic programming is applied to samples along linear ground (as shown in the example of Figure 3.3.23) or image (e.g. epipolar lines) profiles. The larger is the domain, the higher are the chances of finding a 'correct' global solution but the higher is the optimization complexity and the computing times. Besides, if we choose local domains, we have to ensure that all the connected domains are coherent thus leading us to another optimization problem.

The optimization processes can be carried out in the same way in the object space, i.e. on the 3D samples hypotheses or in image space on the matching hypotheses. One of the major interests of these processes besides improving globally the results is that we can use windows of smaller sizes and have a better local and morphological rendering of small 3D structures. Moreover, using the windows that are as small as possible is a guarantee that the neighbouring measurements are the 'more' independent as possible thus lowering the probability of having structured erroneous areas that could mislead the optimization process.

When the surfaces are not continuous but piecewise continuous as in urban areas, the optimization process should take into account the location of the 3D break lines to authorize locally a depth or a disparity discontinuity. As we do not have any a priori information on their 3D location, the only information we can inject in the process is the location of image contours, which are the projections of possible 3D breaks (March, 1989). We can then apply a continuity constraint on the hypotheses (brought back in image space) on each separate side of the contour. As we said earlier, in urban areas, contours corresponding to building limits do not appear in all the images at the same time, but most of the time a given building edge appears contrasted in at least one of the images. Thus if we want the break process to be efficient we have to consider the optimization on all the contours of all the images at the same time. Nevertheless, one should have in mind that this process can in some cases generate virtual depth discontinuities around non-3D contours.

Figure 3.3.24 An example of the management of hidden parts(a): in (b) the left image; (c) the raw DSM, with (in black) the areas where the correlation process failed. Considering that these zones are generally on the ground, by subtracting the DTM (e), obtained by smoothing the data acquired between the raised structures, one gets a DEM (d) with only minor imperfections.

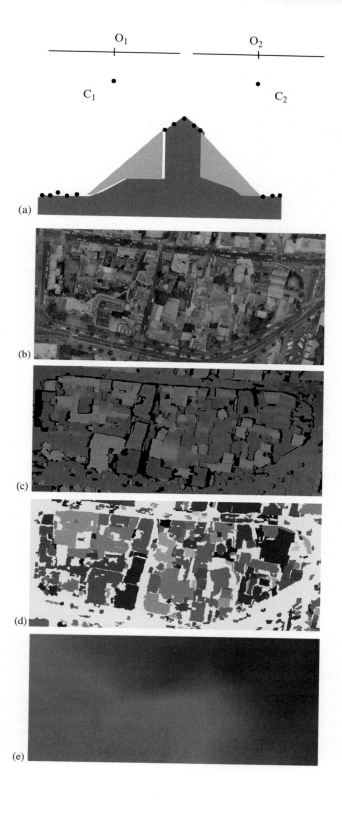

It is necessary to discard, before the optimization process, all the grid points that have no significant correlation profile. Indeed, these points could correspond to areas that are hidden due to occluding structures. Thus looking for a path in these areas would have no sense and would lead in all cases to an erroneous solution. Should these points be a posteriori blindly interpolated in the DEM generation process? Certainly not! This would lead us to odd-looking surfaces. The hole-filling strategy should be decided and guided by the operator. The alternative solution if we desire a fully automatic process would be to have the existence of a solution inside the matching hypotheses for every landscape point. This means that every landscape point should be seen in at least two images of the survey. In the case of urban surveys, this implies a high stereo overlap (along and/or across-track). The optimization process should then be carried out on all the 3D sample hypotheses accumulated from all the possible stereopairs (Roux et al., 1998). Another alternative is to consider that missing areas are on the ground. Identifying and removing all the raised structures through a joint image/analysis will leave inside the DSM only the ground elevation samples. Thus the elevation values for the initial missing areas can be interpolated from all the set of ground elevation samples. (See Figure 3.3.24, colour section.)

3.3.8 An example of strategy for complex urban landscapes: dynamic programming applied to the global and hierarchical matching of conjugate epipolar lines

Olivier Dissard

We will describe here a global matching process based on dynamic programming that integrates contours to model the 3D discontinuities. This scheme (Baillard and Maître, 1999; Baillard and Dissard, 2000) has given very good results on urban, suburban, and rural landscapes.

What is dynamic programming in the case of epipolar images matching?

For the purpose of stereo image matching, dynamic programming should be considered as the search for the 'best' path between two known pairs of homologous points, to derive the 3D profile between these two points. 'The best' is considered in relation to an energy function that assesses the resemblance between each of all the possible pairs of points.

Considering the SMI point of view, epipolar resampled images represent an interesting workspace: considering each pair of homologous lines, the aim becomes to search for a 'best' path between the two origins and ends (or two known pairs of homologous points) in the space defined by the right and left epipolar lines and filled with a similarity function (see Figure 3.3.25). All the explanations that follow are based on a 'pixel-to-pixel' but not a sub-pixel matching.

Similarity function

A necessary condition for the similarity function is that the computation must be comparable everywhere: for example, adaptive windows for the computation of cross-correlation coefficient do not provide comparable values because the height of the correlation peak depends on the size of the window. Some computation tricks reduce the influence of window size without solving it. Thus, in these cases there can not be an objective optimized path.

The similarity assessment used for the matching in Figures 3.3.25 and 3.3.26 is nothing but the simple cross-correlation coefficient on a neighbourhood with constant size. It does not assess the resemblance of the local radiometry, but the resemblance of the radiometry change inside the neighbourhood. We will see later how to take radiometric resemblance into matching with dynamic programming.

Possibilities for paths

On the assumption that objects on both epipolar lines should be found in the same order (which is quite the reality, except mobile objects or perspective effects on lateral sides of the epipolar pair between pylons or trees and ground texture), each possible path that goes through a point comes from only three directions (Figure 3.3.25). It means in fact that there is no possibility of going back on one image when going forward on the

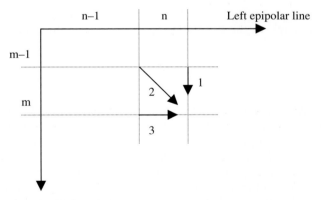

Figure 3.3.25 Possible paths for matching with epipolar lines.

other one. This simple but quite robust assumption makes the computation possible, for the best path through a point derives from three best paths with coordinates less than or equal to the considered ones.

The selected path on the point $P(n, m)$ is then:

$$P(n, m) = \text{Max}\big(P(n, m-1) + \text{Energy}(1), P(n-1, m-1) + \text{Energy}(2), P(n-1, m) + \text{Energy}(3)\big). \qquad (3.16)$$

Let us examine each path:

- Path 2 means that the disparity of (n, m) is the same as the disparity of $(n-1, m-1)$: we are on a horizontal 3D surface.
- Path 1 means that we have moved one pixel forward on the right image, keeping the same position on the left one, disparity has progressed. If (n, m) belongs to the right path, it means that we are on a part of the right image that does not appear on the left one: occlusion, mobile vehicle, specularity, etc.

  Path 1 is also for sloped areas: for example, a ½ sloped profile will alternate Path 1 and Path 2 through the optimized path.
- Path 3 is the opposite of Path 2, regarding left and right lines.

Into the algorithm, point (n, m) will be assigned with the value of $P(n, m)$ and $O(n, m)$ which is the provenance (1, 2 or 3) of $P(n, m)$:

$$O(n, m) = \text{Arg(energy)}\big(\text{Max}(P(n, m-1) + \text{Energy}(1), P(n-1, m-1) + \text{Energy}(2), P(n-1, m) + \text{Energy}(3))\big). \qquad (3.17)$$

Thus P is used for progressing from the beginning point to the end point, while O allows going back to the beginning point, once the end point is reached.

Energy function

Let us consider Path 2: it means that disparity is the same between (n, m) and $(n-1, m-1)$, in other words, it means that the 3D profile is continuous, the homology between points n and m is acceptable, therefore the similarity function must provide an acceptable value.

Considered from an opposite point of view, we will assume that Path 2 is the right one if the similarity function (the correlation coefficient $cc(n, m)$) is acceptable, higher than a given threshold $cc_{\min}$.

$$\text{Path 2} = cc(n, m). \qquad (3.18)$$

What happens when $cc(n, m) < cc_{min}$? '$cc(n, m) < cc_{min}$' means that (n, m) is not a homologous pair, we are on a part of one image that is not visible on the other one. Thus no participation of the similarity function can be considered, therefore a constant energy is attributed to Paths 1 or 3.

Progressing towards the end point with Paths 1 or 3 must not be a handicap, compared with a poor matching (Path 2).

It means that if Path 2 is poor, Path 1 + Path 3 > Path 2. On the other hand, if Path 2 is good, Path 2 > Path 1 + Path 3.

$$\text{Thus Path 1} = \text{Path 3} = \frac{cc_{min}}{2}. \quad (3.19)$$

Application

Figure 3.3.27 shows two epipolar lines, a correlation coefficient matrix, and the optimized path between beginning and end points. This algorithm is very convenient for urban context where one usually has to deal with continuous 3D areas, mixed with disparity gaps due to vertical façades. Mobile vehicles, or opposite contrasts due to specular reflection are overcome with series of 'Path 1 + Path 3'. However, the algorithm is robust, for it finds the right path very quickly after such accidents.

Finally the optimized path is 'symmetric': the paths are the same if we go from 'end' point to 'beginning' point.

There is no necessity in having homologous pairs of points as beginning and end points, because the incidence is in fact very limited; the right path is always directly found after the right number of Paths 1 or Paths 3.

Intervals of search can be limited to reduce computation time, for example with a map of 3D contours: the beginning point is a 3D contour $C_1(n_1, m_1)$ and the end point is the next one $C_2(n_2, m_2)$: the optimized path has a length $L < n_2 - n_1 + m_2 - m_1$.

Forbidden paths: how to introduce new constraints

For different reasons, some pairs of points cannot be matched. It is expressed on the matrix by parts or points, through which every Path 2 has a very low energy: instead of computing the correlation coefficient on these points, we decide that $cc \ll -1$. These pairs of points can, however, belong to an occlusion (Path 1 or 3), not to a matched surface.

An example of applying this assumption is a two-step matching process: the first one is made with dynamic programming, having a very strong cc_{min}, 0.7 or 0.8. It gives a set of homologous points, not dense but sparse, sufficient to model the global relative radiometric behaviour of one image according to the other one. Thus a second matching allows a low cc_{min} (always greater than 0, if not it will allow matching opposite contrasts) with forbidden matching where the two radiometries are far from our

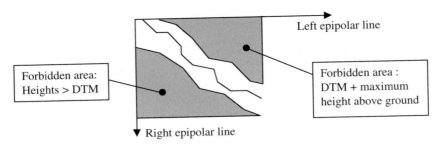

Figure 3.3.26 Restriction of the space search.

Figure 3.3.27 Correlation matrix corresponding to the matching of the two epipolar lines appearing in yellow in the image extracts.

Every grey level with coordinates. *(i, j)* in this matrix corresponds to the correlation score and the likelihood of the matching hypothesis between pixel of row number *i* along the epipolar line in image 1 and pixel of row number *j* along the epipolar line in image 2. The best path found with the dynamic programming within the correlation matrix is superimposed in yellow. The red lines show some examples of corresponding points found along the path. The green lines and areas show, respectively, some examples of 3D occultation contours and the corresponding occluded area in the other image.

radiometric modelling. The examples in Figures 3.3.28 (colour section), 3.3.29 and 3.3.30 are obtained with such a two-step algorithm.

Some parts are forbidden even for occlusion, for example in the case where we know lower and higher possible altitudes (with a coarse matching, or by the knowledge of a DTM and a maximum height of the above-ground objects – trees or buildings). Restriction of the search space greatly speeds up the process (see Figure 3.3.26).

What matching difficulties does dynamic programming solve and not solve?

This algorithm is conceived to solve the matching difficulties due to above-ground vertical façades, and in fact, it solves the problem of mobile vehicles, when no matching ambiguity appears. Repeated textures, such as parking, are also solved by dynamic programming, considered as global optimization.

To the contrary, this algorithm suffers from the similarity measure that takes contrast into account, and thus the borders of 3D objects are un-localized with an error equal to the radius of the measure neighbourhood.

There are still matching ambiguities: a mobile vehicle matched with a parked one – see Figure 3.3.30, confusion with objects on buildings' walls occluded on one image, etc.

One can find more details in Baillard and Dissard (2000). Also, see Figures 3.3.27 and 3.3.28 (colour section).

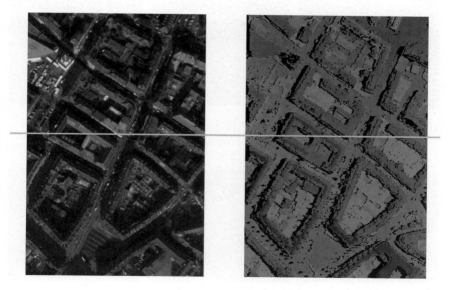

Figure 3.3.28 Results of the global matching strategy on a complete stereopair overlap in a dense urban area. The yellow line superimposed on the two images corresponds to the line we have matched previously in Figure 3.3.27.

3.3.9 Improving the DEM generation in urban areas with higher quality data: multiple views and digital images

Nicolas Paparoditis

At these levels of resolutions and in a context of dense urban tissue, there are a considerable number of hidden parts to the extent that a dense and thorough description of the scene by a stereo analysis is limited (Figure 3.3.29). Besides in urban areas, the classical stereo processing admits some flaws due to all the matching ambiguities encountered, e.g. mobile vehicles (Figure 3.3.30), repetitive structures and also due to the poor image quality. Poor image quality for instance in scanned images is due to noise added by the scanning process, to the dust, the scratches, and other dirt present on the silver halide photographs, but also due to the non-linearity of the silver halide sensibility. Indeed, the signal to noise ratio is dreadful in dark areas such as shadows, which will thus have poor chances of being matched if they are hard.

Digital images

The quality of the images, i.e. the signal to noise ratio, is the key factor for the quality of the products that can be derived from these images.

Figure 3.3.29 Missing information in the DSM due to the hidden areas in the images especially areas around building ground footprints.

Figure 3.3.30 Artefacts due to mobile vehicles. Indeed the stereo-process supposes that the observed surfaces are still between the two acquisitions.

Figure 3.8.4 Scanned images after balancing.

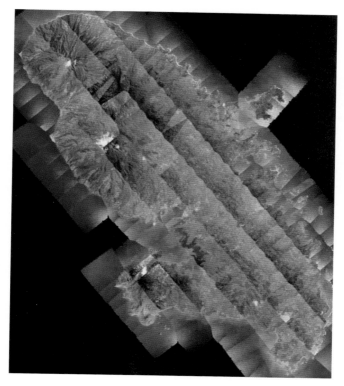

Figure 3.9.2 Example from a real production problem: the radiometric balancing on digitized pictures.

Figure 3.9.4 Examples from the Ariège, France: problems posed by cliffs.

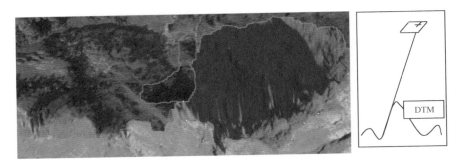

Figure 3.9.5 Generation of stretched pixels.

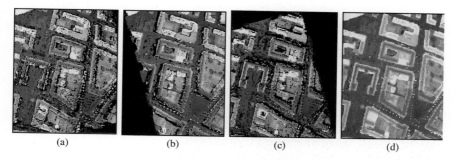

Figure 3.3.34 (a) (b) (c) Three elementary stereopair DSM on the same scene. One can notice that the hidden areas are not located at the same place; (d) DSM obtained by merging the three elementary DSMs.

Figure 3.3.35 Zoomed extracts of the DSMs appearing in Figure 3.3.32.

technique (e.g. based on the median value). This merging allows a densification and reliabilization of the results to a large extent when the elementary DSM are complementary and when the results are valid in most of all the DSM. (See Figures 3.3.34 and 3.3.35.)

The stereo matching process requires the use of non-negligible window sizes if one is looking for sufficiently reliable measures. One knows well that the morphological quality of the DSM, i.e. the ability to render discontinuities, slopes, slope breaks, surface microstructures, all of utmost importance for numerous applications, is directly dependent on the window size. The larger the windows and the more the 'high frequencies' of the surface are smoothed out, the more the depth discontinuities are delocalized, and the more the matching of steep slopes is difficult (due to image deformations). (See Figure 3.3.36.)

One way to reduce the window size is the signal to noise ratio of the images so as to reduce matching ambiguities. This can be obtained naturally by having an imaging system of higher quality, or artificially by increasing the number of observations.

Figure 3.3.36 DSM with superimposition of the real roof building edges. One can see the delocalization of the building depth discontinuities.

Generating the DSM with all the images at the same time: a multi-image matching process guided from object space

A process based on the concept of direct multi-image matching guided from object space can be used to process all the images at the same time to generate the DSM (Figure 3.3.37) (Paparoditis *et al.*, 2000). This process allows the matching of the images and the construction of the raster DSM, and the orthoimage all at the same time without having to go through the processing of all stereopairs and the merging of all elementary DSMs.

The process is based on the following algorithm. For every single node (x, y) of the DSM raster grid, a correlation profile is constructed gathering all the correlation scores calculated for each of the possible z through a plausible interval $[z_{min}, z_{max}]$ depending on an a priori knowledge of the scene (given by a map, a gross DTM, etc.). For each given z, one calculates in the whole set of images, the hypothetical corresponding (i_k, j_k) image coordinates in each image space k (where $0 < k < n$ and n is the number of images where this (x, y, z) point is seen). The likelihood of these hypotheses is given by a direct measurement of the similarity of the set of windows centred on the (i_k, j_k). The 'estimated' z value retained for a given (x, y) is the one for which the optimal value of the correlation profile is reached. Besides, for this (x, y) and for this estimated z we can calculate directly the corresponding grey level in the orthoimage from the set of associated (i_k, j_k) grey levels by taking for instance the average or the median of the values.

A multi-image similarity function

How can one generalize the similarity function to a set of image windows? By defining for example a criterion describing the collinearity dispersion

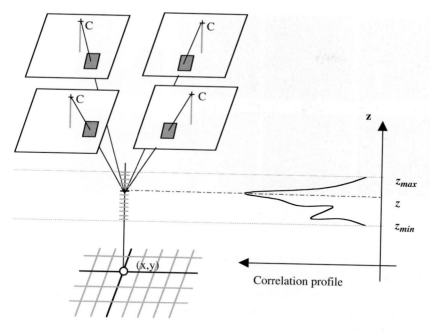

Figure 3.3.37 Multi-image matching guided from object space.

of all texture vectors corresponding to all template windows. One can do this by using this new multi-image texture similarity (MITS) function:

$$0 \leqslant \text{MITS}(i_1, j_1, i_2, j_2, \ldots, i_n, j_n) = \frac{\text{Var}\left(\sum_{k=1}^{n} \mathbf{V}_k(i_k, j_k)\right)}{\sum_{k=1}^{n} \text{Var}(\mathbf{V}_k(i_k, j_k))} \leqslant n, \quad (3.20)$$

where $\mathbf{V}_k(i_k, j_k)$ is the texture vector associated to the window centred on the pixel (i_k, j_k) in image k and $\text{Var}(\mathbf{V}_k(i_k, j_k))$ the variance of the texture vector $\mathbf{V}_k(i_k, j_k)$, i.e. the grey levels inside the window centred on the pixel (i_k, j_k) in image k. Why this correlation function? If the image texture windows are alike, i.e. the texture vectors are collinear, the similarity score is maximal.

A radiometric similarity weighting

In view of the great radiometric stability of digital images (which is not the case of scanned images) and under the assumption that most of the objects that describe the landscape have Lambertian characteristics, we

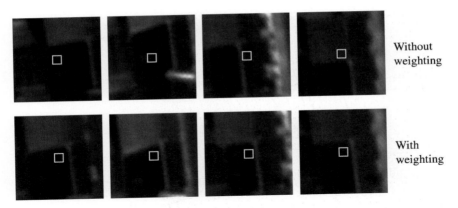

Figure 3.3.38 Comparison of homologous windows (3×3) found for a given (x, y) point without and with the radiometric weighting function.

impose an additional constraint on the absolute radiometry of homologous neighbourhoods, in the shape of a weighting function. This weighting is necessary. Indeed, the MITS similarity function is 'centred average'. This function measures the similarity of the textures of all neighbourhoods but not their radiometric similarity; this can give rise to mismatches. Our new similarity function called MITRAS (multi-image texture and radiometric similarity) can be expressed in the following way:

$$\text{MITRAS}\,(i_1, j_1, \ldots, i_n, j_n) = \\ \text{MITS}\,(i_1, j_1, \ldots, i_n, j_n) \exp -\left(\frac{\text{Var}_k(I_k(i_k, j_k))}{k}\right), \qquad (3.21)$$

where: Var is the variance, $I_m(i, j)$ the grey level of pixel (i, j) in image I_m, and k a normalizing coefficient. The application of this weighting function permits a similarity criterion to be obtained which characterizes at the same time the texture and the radiometric resemblance of the neighbourhoods. Therefore the MITRAS similarity function is more robust and discriminating (Figure 3.3.38).

Multiple view DSM results and window size impact

The increase in the number of observations allows on the one hand a sizable reduction of matching ambiguities met with in the classical stereo-processing and thus to increase the reliability of the process. On the other hand, it allows, also thanks to the high quality of digital images, sizes for the correlation windows of 3×3 to be used (Figure 3.3.39). The results show clearly a dense DSM, an extremely good morphological rendering: all relief slopes, slope breaks, discontinuities and microstructures, and a

Figure 3.3.39 DSM generated with 4 images and 3×3 windows.

Figure 3.3.40 DSM generated with 4 images and 5×5 windows.

good localization of the discontinuities (especially the building edges). However, the very small size of the windows is at the origin of false matches in the very homogeneous areas that do not appear with a 5×5 window (Figure 3.3.40). To the contrary, a 5×5 window has a less accurate rendering of microstructures and discontinuities.

Working from object space offers a good number of advantages. One can work in a transparent way on images of different resolutions, and the parameters are expressed in a metric form. Also we avoid the blind resampling of all the 3D samples corresponding to all image matches in order to generate a regular grid DSM in object space. Indeed, these samples when the matching process is guided from image space follow, although their distribution is regular in image space, an irregular spatial distribution in object space. Finally, its major advantage is that it allows, on the

Figure 3.3.41 Orthoimage corresponding to the DSM of Figure 3.3.40. Images with 40 cm ground pixel size from the digital camera of IGN-F on the French town of Le Mans with a 50 per cent across and along track overlap.

Figure 3.3.42 2,000 × 2,000 Raw DSM obtained with our process on 4 images and 3 × 3 windows.

one hand N images to be treated in a natural and transparent way, and on the other hand, the DSM and the corresponding orthoimage to be constructed at the same time.

Matching self-evaluation and automatic filtering of false matches

As expected, the multi-image correlation score does not supply a probability, but a good self-indicator of the measurement reliability. The results (in Figure 3.3.42 and Figure 3.3.43) show that DSM aberrant points have much weaker correlation scores.

A simple experiment of analysis of the correlation score distributions shows this very clearly. Let us consider real and aberrant score distributions. We call real scores those found with our process, and aberrant scores those obtained by the same process with the proviso that one deliberately stands within an altimetric search interval of the same amplitude but containing no relief. One can observe that the two distributions are

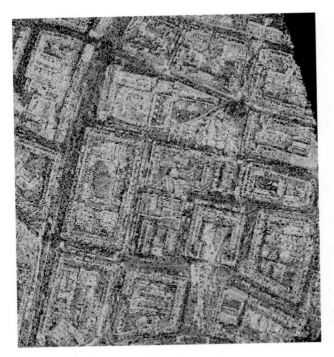

Figure 3.3.43 Correlation scores corresponding to Figure 3.3.42.

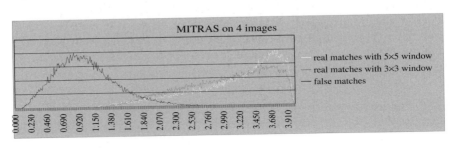

Figure 3.3.44 Correlation score histograms.

but slightly mixed (Figure 3.3.42). From an inspection of the curves, it can be said that on this scene the points having correlation scores above 2.9 are almost 100 per cent sure. (See Figure 3.3.44.)

The relative distribution separation in the case of multi-image matching allows a valid criterion for false points filtering to be defined. It can be noticed that the greater the number of images, the greater is the separation of the two distributions. In the case of stereo-processing, the mixing

Figure 3.3.45 DSM of Figure 3.3.42 filtered with a 2.9 threshold on the correlation scores of Figure 3.3.42.

of the distributions is considerable. This explains the difficulty and the impossibility of defining a reliability criterion and a satisfying rejection threshold on the correlation score values. A low threshold retains a great number of aberrant points, while a high one rejects a very great number of valid points. (See Figure 3.3.45.)

This matching process tends voluntarily to determine solely the landscape points seen in a rather similar way on all the images. As a result, the filtering process will remove all other points and leave missing areas inside the DSM. For example surface points that are only seen in a subset of images will be removed. Nevertheless, we have here the first step of a general matching strategy that can be to complete in a progressive and hierarchical way the DSM starting from the most reliable measures.

3.3.10 Generating a DTM from a DEM: removing buildings, bridges, vegetation, etc.

Nicolas Paparoditis

In the case of dense urban areas and extended areas of vegetation, even if the stereo-process is carried out on SPOT like low-resolution images, the measured elevation samples will describe the rooftops and the canopy surface. The only proper way of generating a DTM from a DEM is by removing from the DEM all the samples corresponding to 3D raised structures, i.e. vegetation, buildings, etc. and fitting surfaces through the holes. This process requires a 3D analysis of the DEM, which can be combined with an image analysis to detect and recognize all these cartographic features.

References

Baillard C., Dissard O. (2000) A stereo matching algorithm for urban digital models, *Photogrammetric Engineering and Remote Sensing*, vol. 66, no. 9, September, pp. 1119–1128.

Baillard C., Maître H. (1999) 3D reconstruction of urban scenes from aerial stereo imagery: a focusing strategy, *Computer Vision and Image Understanding*, vol. 76, no. 3, December, pp. 244–258.

Bertero M., Poggio T., Torre V. (1988) Ill-posed problems in early vision, *IEEE Proceedings*, vol. 76, no. 8, pp. 869–889.

Canu D., Ayache N., Sirat J.A. (1995) Accurate and robust stereovision with a large number of aerial images, *SPIE (Paris)*, vol. 2579, pp. 152–160.

Cord M., Paparoditis N., Jordan M. (1998) Dense, reliable and depth discontinuity preserving DEM computation from H.R.V. images, *Proceedings of ISPRS Commission II Symposium, Data Integration: Systems and Techniques*, Cambridge, England, July, vol. 32, no. 2, pp. 49–56.

Gabet L., Giraudon G., Renouard L. (1994) Construction automatique de modèles numériques de terrain à haute résolution, *Bulletin SFPT*, no. 135, pp. 9–25.

Grimson W. (1983) An implementation of a computational theory of visual surface interpretation, *Computer Vision Graphics Image Processing*, no. 22, pp. 39–69.

March R. (1989) A regularization model for stereo vision with controlled continuity. *Pattern Recognition Letters*, no. 10, pp. 259–263.

Okotumi M., Kanade T. (1993) Multiple Baseline Stereo, *IEEE Transactions on Pattern Analysis and Machine Intelligence*, vol. 15, no. 4, April, pp. 353–363.

Paparoditis N., Cord M., Jordan M., Cocquerez J.P. (1998) Building detection and reconstruction from mid to high resolution images, *Computer Vision and Image Understanding*, 72(2), November, pp. 122–142.

Roux M., Leloglu U.M., Maître H. (1998) Dense urban DEM with three or more high resolution aerial images, *ISPRS Symposium on GIS, Between Vision and Applications*, Stuttgart, vol. 32, no. 4, September, pp. 347–352.

Tikhonov A.N. (1963) Solution of incorrectly formulated problems and the regularization method, *Soviet Mathematical Document*, no. 4, pp. 1035–1038.

3.4 FROM THE DIGITAL SURFACE MODEL (DSM) TO THE DIGITAL TERRAIN MODEL (DTM)

Olivier Jamet

As soon as one is interested in the topographic data production with a rms accuracy in altitude of the order of a metre or less, the derivation of a DTM from the calculated DSM, by manual edition or automatic levelling of the raised structures, is an indispensable step. The semi-automatic processes permitting one to alleviate the edition of the DSM can be classified in three categories.

3.4.1 Filtering of the DSM

Techniques of filtering are based on the hypothesis that elements of the raised structures constitute connex zones of limited area and presenting strong contrasts of elevation with their environment. The method most often used is issued from mathematics morphology: the DTM is obtained by morphological opening of the DSM by a structuring element defined by its size and its geometry (parameters of the filter). This method is perfectly rigorous only in a plane land. Indeed it doesn't preserve the curvature of the land. The quality of the DTM produced will be therefore worse if the land is hilly and if the structuring element is of large size. Besides, the abrupt irregularities of the land leads to artefacts that can become very visible for openings of large size, as in the case of the elimination of building envelopes in a metric resolution DSM, for example. These artefacts can be accentuated by the noise of the DSM, and more precisely by the presence of local parasitic minima. A filtering of these local minima, for example by an applied morphological closing of the DSM, will then improve the results.

A variant of this method of levelling of the raised structures consists in applying openings of increasing successive sizes with structuring elements of variable shape up to the wanted size. If this method doesn't preserve the accuracy of the DTM better in the presence of strong curvatures, it has the advantage of producing a smoother result.

3.4.2 Segmentation and elimination of the raised structures

An alternative to the simple filtering consists in using techniques of shape recognition to circumscribe the extension of the raised structures. The DTM is then calculated by replacing, on the extension of the detected raised structures, the initial elevation values by interpolated values. This segmentation of the raised structures can be done by a process of the DSM only, or by analysing jointly an orthophotography of the same zone.

3.4.2.1 Segmentation of the DSM

All tools of segmentation used in picture analysis can obviously be used for the analysis of the DSM, requiring a definition of the criteria characterizing the raised structures in relation to the ground.

The simplest methods are based on the examination of slopes or curvatures of the DSM for a detection analogous to a detection of contour in an image. The selection of the raised structures in the extracted contours is made by imposing that contours are closed and that the value of the slope on contours is greater than a minimal value. A supplementary constraint on the size of the detected zones can also be applied. These techniques are not recommended. They will always be extremely sensitive to the accuracy of the DSM at the edge of the raised structures, though this accuracy is impossible to guarantee (either because of qualities of the picture having served to the extraction, or because of the hidden parts in the case of a matching on a couple of pictures).

It appears therefore preferable, if one uses only elementary operations of image process, to use the method described in §3.4.1. A simple threshold on the difference between the DSM and the DTM produced by morphological opening will indeed give a segmentation of the raised structures a priori more reliable than techniques of contour extraction – this segmentation permitting the calculation of a new DTM by initial data interpolation.

To palliate the fragility of methods based on contour extraction on the DSM and the shortcomings of methods based on mathematical morphology, which will fail systematically in cases of extended raised structures (islets of buildings in urban zone, forests, etc.) and hilly landscapes, some more elaborate methods are made the object of research. In Baillard (1997), the raised structures is defined by a Markovian model in which the labelling of a region as raised structures is a function of relations of dependence with its neighbouring regions (e.g. 'a zone higher than a neighbouring zone recognized as belonging to the raised structures necessarily belongs to the raised structures'). Methods of segmentation by area growth, with a criterion of homogeneity based on differences of altitude between neighbouring points are proposed also in Masaharu and Hasegawa (2000), and Gamba and Casella (2000) for example. Other authors use some cartographic data to force the segmentation (e.g. Horigushi et al. (2000)). (See Figure 3.4.1, colour section.)

3.4.2.2 Segmentation of an orthophotography superimposable on the DSM

The interpretation of photographic pictures of the studied zone, if not alone able to distinguish the raised structures without ambiguity, can improve the reliability of processes of segmentation of the DSM. In particular, specific

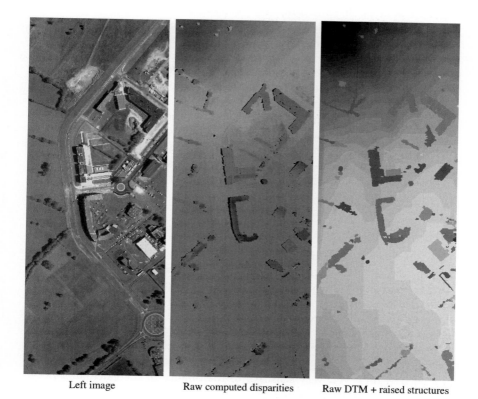

Left image   Raw computed disparities   Raw DTM + raised structures

Figure 3.4.1 Images from the digital camera of IGN-F (1997) on Le Mans (France). The outstanding signal/noise ratio allows one to get excellent matching results on the automatic correlation process, particularly on homogeneous or shadowy zones where normal digitized images always produce many difficulties. Benefit is taken from this situation, by using small correlation windows (5×5 and even 3×3), which improves the planimetric accuracy.

detectors (recognition of the structure, vegetation) applied on an orthophotography superimposable on the DSM will provide complementary masks combinable with the segmentation of the DSM. Dissard applies this technique, for example, for the detection of forest zones.

3.4.3 Selection of ground points in the DSM

A third approach consists in selecting, by an ad hoc procedure, a set of points situated on the ground in the DSM, and rebuilding the DTM by interpolation of this set of points. This technique has the advantage of avoiding the problem of an exact segmentation of the extensions of the raised structures. It can, however, be used only in the individual cases where the strategy of selection of ground points is very reliable (the sensitivity to errors being as large as the density of the selected seeding is low).

It is, for example, the case in urban zones, when the land varies little. One can be content with a scattered set of points, which one can choose, for example, on the road network (if a cartography is available, or while calculating an automatic extraction).

In the absence of other information, it is also possible to use procedures of blind selection of local minima in an analysis procedure of the DSM by mobile window, the window of analysis being chosen of sufficient size not to be ever entirely covered by the raised structures. This method, however, is rigorous only in flat landscapes and is extremely sensitive to the noises of the DSM.

3.4.4 Conclusion

These methods cannot be considered as perfectly operative, however.

Only the simplest algorithms (as algorithms of filtering that are most current) are generally present on digital photogrammetric workstations. They will give most of the time good results on landscapes of moderate slope and when the raised structures is constituted of connex parts of weak extension. They remain therefore precious from the moment the number of these connex parts is high, because they will allow one to save a lot of editing time.

They are, however, very sensitive to the quality of the calculated DSM, and in particular to the slope of the DSM at the borders of the raised structures (that must be the strongest possible). They can also produce some unexpected results on ambiguous superstructures such as bridges, which can either be considered as belonging to the raised structures or not. They must not therefore be used without supervision.

References

Baillard C. (1997) Analyse d'images aériennes stéréoscopiques pour la restitution 3D des milieux urbains: Détection et caractérisation du sursol. Ph.D. Thesis, École Nationale Supérieure des Télécommunications, Paris.

Horigushi S., Ozawa S., Nagai S., Sugiyama K. (2000) Reconstructing road and block from DEM in urban area, Proc. ISPRS Congress 2000, *International Archives of Photogrammetry and Remote Sensing*, vol. XXXIII, Part B3, pp. 413–420.

Gamba P., Casella V. (2000) Model independent object extraction from digital surface models, Proc. ISPRS Congress 2000, *International Archives of Photogrammetry and Remote Sensing*, vol. XXXIII, Part B3, pp. 312–319.

Masaharu H., Hasegawa H. (2000) Three-dimensional city modelling from laser scanner data by extracting building polygons using region segmentation method, Proc. ISPRS Congress 2000, *International Archives of Photogrammetry and Remote Sensing*, vol. XXXIII, Part B3, pp. 556–562.

3.5 DSM RECONSTRUCTION

Grégoire Maillet, Patrick Julien, Nicolas Paparoditis

3.5.1 Digital elevation models under TIN shape

Grégoire Maillet

The natural form of the DEM is raster (a regular grid of altitudes), either acquired by correlation or by laser scanning. But, given their extremely heterogeneous nature and the presence of discontinuities, this format is neither adapted to their storage nor to their manipulation. The construction of a TIN (for triangulated irregular network) presents the two favourable features to reduce greatly the volume of data (a reduction of around 90 per cent for an urban DEM) and to be more effective for all applications of visualizations [1] (Figure 3.5.1.5). Besides, this construction is a first raw data interpretation that may facilitate the analysis of the scene [2].

Contrary to the general problems of surface triangulation, the raster origin of data makes it possible here to work in 2.5D. Most of the existing methods are based on the recursive insertion of points in a planimetric triangulation up to the satisfaction of a criterion on the number of points or on the quality of the approximation [3]. The problem can be decomposed therefore into two independent sub-problems: the way to triangulate points and the criterion of the choice of the points to insert.

A triangulation by a simple splitting into three of the triangle to the vertical of the point to insert leads inevitably to triangles of very acute shape that are very difficult to manipulate. The criterion of Delaunay allows the appearance of these problems to be minimized, and there are very efficient implementations of insertion of points in a Delaunay triangulation [4][3]. Nevertheless, the lack of taking into account the altitude can lead to optimal triangulations in planimetry, but disjointed in 3D (Figure 3.5.1.1).

Different methods exist to take into account the altimetric consistency of the triangulation [3] (Figures 3.5.1.2 and 3.5.1.3), but the altimetric improvement is often performed to the detriment of the planimetric quality of the triangulation. The choice of the points to insert is based in general on a measure of importance, the simplest of these measures being the local error on the considered point, that is to say the gap in Z between the TIN and the raster model. It is a simple and fast measure [3] but one that remains quite sensitive to the noise of the DEM and provides not very meaningful results to the neighbourhood of altimetric discontinuities.

A more robust measure consists in evaluating the volume added by the insertion of a point while taking the volume between the point to insert and the triangle that is on its vertical. This measure will give some more meaningful results in the case of light delocalizations on the sides of the buildings (Figure 3.5.1.4).

At the time of the construction of a TIN by iterative insertion many superfluous points are inserted. They are judged optimal to one given

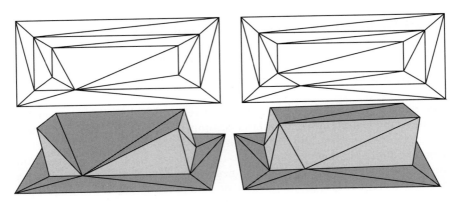

Figure 3.5.1.1 On the left the triangulation of a building and its perspective representation with the criterion of Delaunay, on the right a triangulation taking into account the altitudes.

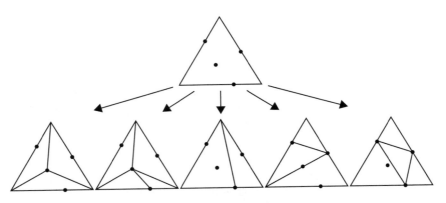

Figure 3.5.1.2 At the time of the insertion of a point there is division of the triangle into three. For each neighbouring quadrilateral the two diagonals are tested. And the solution creating fewer errors in Z is preserved.

instant and permit a convergence toward a good solution but become redundant in the final result. The simplification of surface is therefore a complementary and necessary approach for the construction of an optimal TIN, that is to say describing in the best way the DEM with a minimum of points. This simplification can be performed in a way similar to the insertion: with a method of suppression of points in the triangulation and a measure of the importance of points. But the simplification of TIN is a problem very often met in image synthesis and thus specific methods have been invented. The most frequent approach is not based on the suppression of points, but on the contraction of a pair of points. That is to say

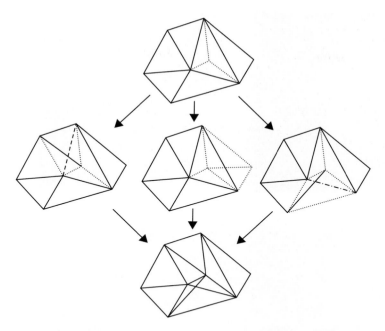

Figure 3.5.1.3 For every triangle one looks for four characteristic points (one in the triangle and one on each side). According to the result of this research one applies one of the five possible divisions.

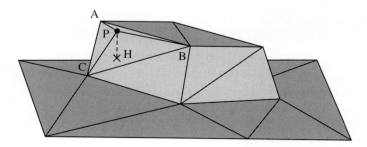

Figure 3.5.1.4 For a point P on a side of roof slightly delocalized the local error PH is large whereas the volume of the ABCP tetrahedron is small.

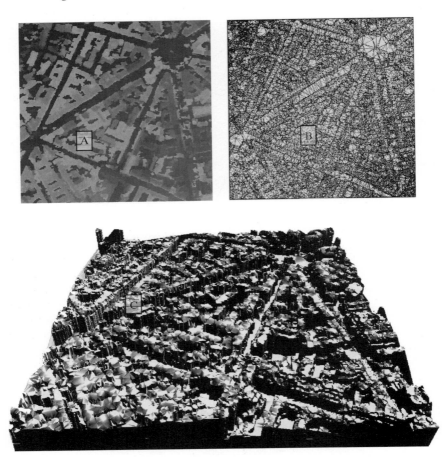

Figure 3.5.1.5 (a) urban DEM obtained by correlation; (b) transformation into a TIN containing 5 per cent from initial points; (c) perspective visualization of the TIN.

the replacement of a couple of points by a new point minimizing the committed error [5].

References

[1] Haala N., Brenner C. Virtual city models from laser altimeter and 2D map data. *Photogrammetric Engineering and Remote Sensing*, vol. 65, no. 7, July 1999, pp. 787–795.

[2] Maas H.-G., Vosselman G. Two algorithms for extraction building models from raw laser altimetry data. *ISPRS Journal of Photogrammetry and Remote Sensing*, vol. 54, nos. 2–3, July 1999, pp. 153–163.

[3] Garland M., Heckbert P.S. Fast polygonal approximation of terrains and height fields. *Technical Report CMU-CS-95-181, Comp. Sci. Dept, Carnegie Mellon University*, September 1995, pp. 1–37.
http://www.cs.cmu.edu/~garland/scape
[4] Devillers O. Improved incremental randomized Delaunay triangulation. *INRIA Research Report no. 3298*, November 1997, pp. 1–21.
[5] Garland M., Heckbert P.S. Surfig. Simplification using quadric error metrics. *SIGGRAPH 97 Proc.*, August 1997, pp. 209–216.
http://www.cs.cmu.edu/~garland/quadrics

3.5.2 Constitution of a DTM in the shape of a regular grid of altitudes, from contour lines in vector mode

Patrick Julien

3.5.2 Position of the problem. Notion of topographic surface

Suppose one is given a file containing a system of contour lines describing the relief of a fixed geographical zone. To simplify, we call a contour line what is in fact a section of a contour line, either closed, or stopped on the edge of the zone. Every contour is assumed to be represented in vector mode, i.e. described by vertices of a polygonal line, or by an ordered sequence of points, say $\{p_{ij} = (x_{ij}, y_{ij}, z_j); i = 1, \ldots, n(j)\}$ for the contour

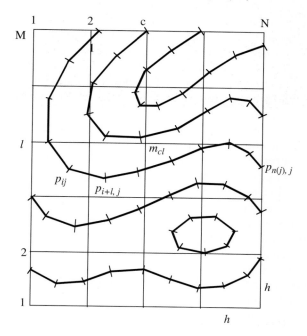

Figure 3.5.2.1 Contour lines in vector mode.

line placed to the rank j of the file; the coordinate z is the altitude, and the x, y coordinates are relative to a cartographic representation. It is important to point out that the contour line is not only the finite family of points p_{ij}, but the union of segments $[p_{ij}, p_{i+1,j}]$; it is therefore a continuous series, whose family p_{ij} is one of the possible representations; in particular, one is allowed to subdivide any segment $[p_{ij}, p_{i+1,j}]$ into smaller segments, which doesn't change the shape of the curve. (See Figure 3.5.2.1.)

Now, one wishes to 'interpolate' in these contour lines a grid of points with a grid interval h:

$$\{m_{cl} = (x_c, y_l, z_{cl}); \quad c = 1, \ldots, N, l = 1, \ldots, M\}, \tag{3.22}$$

where $x_c = ch$, $y_l = lh$, regular in the x, y coordinates system. By 'interpolating', we mean that every point must be situated on 'the' surface represented by the contour lines, which surface is in principle well defined by the rule of surveying these contours, according to which rule their plot should be such that the user can always consider the slope between two regular or intermediate curves as regular.

For example, the rule above means that if the user wants to draw a new contour line situated at half height of two successive ones, he must place it at an equal horizontal distance from the first ones, for if it were not the case the slope would be steepest on one side of the new contour than on the other, and therefore non-regular; more generally any intermediate contour must be plotted between the two neighbouring contours, with horizontal distances in ratio to vertical distances. In a more precise way, we will interpret the rule while saying that, between two successive contours and along a steepest slope line, the scalar value of the slope can be considered constant. One so rejoins the rule according to which contour lines must permit us to determine, with a sufficient precision, the elevations of any ground point by simple linear interpolation.

We insist on the fact that it is necessary to admit that these rules give place, for several users, to negligible interpretation differences, in other words that contour lines suggest to all users the same 'topographic' surface (simplified model of the real ground surface), obtained in practice by the rule of the linear interpolation next along the line of steepest slope; it is on this surface that the grid points m_{cl} must lie.

So the interpolation of points m_{cl} is transformed into the plotting of the steepest slope lines, which are the orthogonal lines to contours (Figure 3.5.2.2). However, the tracing of these orthogonal lines, that essentially consists in joining between each other bootjacks placed on curves, require, as soon as orthogonal lines are a little wavy, a certain manual ability. One guesses that the algorithmic transposition of this ability is delicate, and the methods presented below only achieve it in an approximate way.

We are going to describe here methods of approximation by triangular

Figure 3.5.2.2 Handmade drawing of the steepest slope lines, orthogonal to contours.

irregular network surface, by thin plate spline surface, and by elastic grid surface.

3.5.2.2 Approximation of the topographic surface by a triangular irregular network surface (TIN surface)

Definitions relative to triangulations

Let's first specify the vocabulary used. A triangulation of a domain D of the plane is a finite family of triangles of which union is D, of non-null areas, of which any two have in common either nothing, or a vertex, or a side. A triangulation is called Delaunay's if the open circle (i.e. a circle without its edge) circumscribed to any triangle contains no other triangle vertex. One shows that, given a finite family $\{m_i\}$ of points of the plane, there exists a triangulation of Delaunay, and algorithms to construct it, whose vertices of triangles are exactly the points m_i; the triangulation is unique if there are never four cocyclic points m_i, m_j, m_h, m_k. A triangulated surface is a connex surface, union of a family of (plane) triangles of the space, of non-null areas, of which any two have in common either nothing, or a vertex, or a side.

Conditions to be observed by a TIN surface approximating the topographic surface

We have now to approximate the topographic surface with a triangulated surface fitted on contour lines. The searched-for surface here is with irregular faces, and the horizontal positions of triangle vertices are not a priori known; this problem is therefore distinct from fitting a surface whose horizontal positions of vertices would form a fixed in advance network, a problem in which only the altitudes of triangle vertices are unknown.

It seems natural that triangles' vertices are points taken on contour lines, because there doesn't appear a simple way to define other points for this use. Besides, every triangle must be an acceptable approximation of the topographic surface, which requires that its three sides are close to this surface. That in particular notably forbids that a triangle edge 'crosses' a contour (in the sense that their horizontal projections intersect each other), because, in general, such an edge no longer lies on the topographic surface; for example, on Figure 3.5.2.3, the AB side crossing the contour $(z + e)$ does not lie on the topographic surface whose profile is AIB. In particular, a triangle cannot have its vertices on three contours of different altitudes, otherwise one of its sides would 'cross' a contour (ABC triangle). A triangle must not have its three vertices on the same curve, otherwise it would be horizontal and would not lie on the topographic surface (DEF triangle). It is therefore necessary that every triangle has its vertices on two consecutive contours; in addition, the side defined by the two vertices placed on the same contour must be confounded with the contour (GHJ triangle) for, if it were not the case, this side could not lie on the surface.

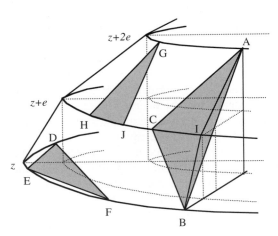

Figure 3.5.2.3 ABC is an unauthorized triangle because it touches three contours; DEF is unauthorized because it touches only one contour; GHJ is an authorized triangle, touching two consecutive contours, with one side along a contour.

It is necessary to also observe, for example on the GHJ triangle, that the smaller the HJ side in relation to the sides GH and GJ, the better is the approximation of the topographic surface; it is therefore a priori preferable that sides bordering the curves be short.

It is finally necessary that triangles overlap the whole zone described by contour lines, so that every point of any contour line is on a side of a triangle; and that above every horizontal position (x, y) there is only one triangle. In these conditions, the horizontal projections of triangles constitute a plane triangulation. Once the future triangle vertices (X_i, Y_i, Z_i) are chosen, one will therefore get a triangulated surface by constructing a plane triangulation whose vertices of triangles are points (X_i, Y_i).

Construction of a triangulated surface

Among all possible plane triangulations, that of Delaunay seems a reasonable choice, because its geometric properties are known, unlike a triangulation that would be constructed with the help of heuristics.

It remains to choose the points devoted to be the triangles' vertices. It is necessary to take at least all points $p_{i,j}$ initially defining the contour lines (except possibly those aligned with their predecessor $p_{i-1,j}$ and their successor $p_{i+1,j}$, which doesn't modify the tracing of the contour). However, if the distance between two consecutive points $p_{i,j}, p_{i+1,j}$ is large in relation to their horizontal distances d_1, d_2 to neighbouring curves, the triangulation of Delaunay risks the construction of triangles crossing a curve, therefore forbidden.

For example from the four points p, p', q, r of Figures 3.5.2.4(a) and (b), one can construct either the authorized pair of triangles $\{pp'q, pp'r\}$, or the pair $\{pqr, p'qr\}$, forbidden because the side pq 'crosses' a contour.

In the case where angle (q) + angle $(r) < \pi <$ angle (p) + angle (p') (Figure 3.5.2.4(a)), the pair of triangles is compatible with the condition of Delaunay because neither r is in the $pp'q$ circle, nor q in the $pp'r$ circle, whereas the $\{pqr, p'qr\}$ pair is incompatible since p is in the $p'qr$ circle and p' in the pqr circle. In the opposite case where angle (p) + angle (p') $< \pi <$ angle (q) + angle (r) (Figure 3.5.2.4(b)), the pair $\{pqr, p'qr\}$ is Delaunay compatible, but not the pair $\{pp'q, pp'r\}$.

So in order that the Delaunay triangulation constructs the acceptable pair, it is necessary to be in the case: angle (q) + angle $(r) < \pi$; as angle $(q) < 2$ Arctan $(d/2d_1)$ and angle $(r) < 2$ Arctan $(d/2d_2)$, it is sufficient that:

$$\text{Arctan}\left(\frac{d}{2d_1}\right) + \text{Arctan}\left(\frac{d}{2d_2}\right) < \frac{\pi}{2} = \text{Arctan}\left(\frac{d}{2d_1}\right) + \text{Arctan}\left(\frac{2d_1}{d}\right),$$

i.e. $d < 2\sqrt{d_1 d_2}$. (3.23)

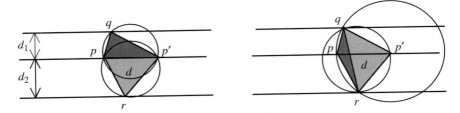

Figure 3.5.2.4 (a) (left) Topographically authorized $pp'q$ and $pp'r$ triangles are selected by Delaunay triangulation; (right) Topographically unauthorized triangles pqr and $p'qr$ are rejected by Delaunay triangulation.

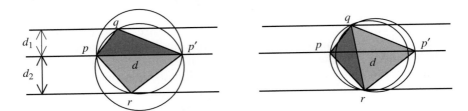

Figure 3.5.2.4 (b) (left) Triangles $pq'q$ and $pp'r$ are topographically authorized, but rejected by Delaunay triangulation; (right) Triangles pqr and $p'qr$ are topographically unauthorized, but selected by Delaunay triangulation.

If d does not verify this condition, additional points between p and p' are necessary, at mutual distances $d' < 2\sqrt{d_1 d_2}$; the triangulation of Delaunay then constructs authorized triangles.

The problem of horizontal triangles

There is on the other hand an unavoidable situation, that is the presence of triangles whose three vertices are on the same curve, i.e. horizontal triangles, in thalwegs and on crests. One sees indeed on Figure 3.5.2.5 that a forbidden horizontal triangle such as $pp'p''$ respects the condition of Delaunay, whereas an authorized triangle such as $pp'q$ doesn't respect the condition.

A possible remedy consists in attaching to contour lines some new polygonal line as $qq_1 q_2 \ldots q_m r$ cutting the horizontal triangles (Figure 3.5.2.6), with altitudes $q_1, q_2, \ldots, q_m$ linearly interpolated according to the curvilinear abscissa on qr (the line qr can be seen as an approximation of the crest or thalweg line).

DSM reconstruction 235

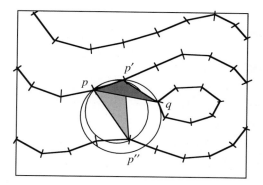

Figure 3.5.2.5 Topographically authorized triangle $pp'q$ is rejected by Delaunay triangulation; instead the horizontal unauthorized triangle $pp'p''$ is selected.

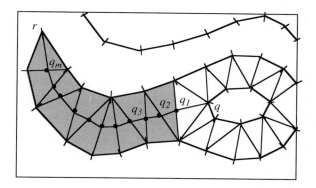

Figure 3.5.2.6 From the horizontal triangles (in grey) selected by Delaunay triangulation, a polygonal line $qq_1 q_2 \ldots q_m$ is defined, with each q_i in the middle of an edge, and: Altitude (q) > altitude (q_1) > altitude (q_2) > $\ldots$ > altitude (q_m) > altitude (r).

One constructs then a new Delaunay triangulation (Figure 3.5.2.7), if necessary by adding some intermediate points on segments $[q_1, q_2]$, $[q_2, q_3], \ldots, [q_{m-1}, q_m]$, to guarantee that no triangle rides the line qr. Then the TIN surface thus constructed is a relatively satisfactory approximation of the topographic surface.

Going from TIN surfaces to the regular grid of altitudes

Once the surface with triangular faces is obtained, it remains to determine in which triangle of the triangulation each m_{cl} point is projected, to form

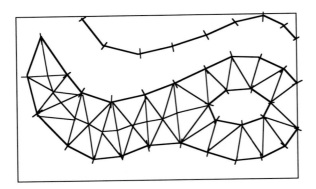

Figure 3.5.2.7 A part of the improved Delaunay triangulation (including the additional vertices $q\ q_1\ q_2 \ldots q_m$) with topographically authorized triangles.

the equation $z = ax + by + d$ of the plane of this triangle, and to calculate the elevation $z_{cl} = ax_c + by_l + d$.

3.5.2.3 Approximation of the topographic surface by a thin plate spline surface

Definition

Thin plate spline surfaces answer the following problem of interpolation: given a sample of n points of the plane $m_i = (x_i, y_i)$ and their elevation z_i, find a function $Z(x, y)$ so that $Z(x_i, y_i) = z_i$ for $i = 1, \ldots, n$ and that the integral

$$K(Z) = \int_{R \times R} Z''_{xx}(x, y)^2 + 2Z''_{xy}(x, y)^2 + Z''_{yy}(x, y)^2 \, dx\, dy \qquad (3.24)$$

be minimal; Z''_{xx}, Z''_{xy}, Z''_{yy} designate partial second derivatives.

Thin plate splines also answer the adjustment problem: besides, given n weight values $\mu_i > 0$, find a function $Z(x, y)$ such that the quantity:

$$E(Z) = K(Z) + \sum_{i=1}^{n} \mu_i \left(Z(x_i, y_i) - z_i \right)^2 \qquad (3.25)$$

be minimal.

J. Duchon showed (1976) that each of these problems admits a solution and only one (under the condition – always true – that the sample counts at least three non-aligned points). He also characterized solutions as functions in the shape:

$$Z(x, y) = \sum_i a_i \gamma(r_i(x, y)) + b_0 + b_1 x + b_2 y, \tag{3.26}$$

with

$$\sum_i a_i = 0, \quad \sum_i a_i x_i = 0, \quad \sum_i a_i y_i = 0,$$

where

$$r_i(x, y) = \sqrt{(x - x_i)^2 + (y - y_i)^2}, \quad \gamma(r) = r^2 \ln r.$$

Note that $Z(x, y)$ is defined by continuity at (x_i, y_i) points since $\gamma(r)$ tends toward 0 when r tends toward 0.

More precisely, the solution of the interpolation problem is the function $Z(x, y)$ of the shape Eqn (3.26) such that $Z(x_j, y_j) = Z_j (j = 1 \ldots n)$; i.e. whose $n+3$ coefficients $a_1, \ldots, a_n, b_0, b_1, b_2$ verify the $n+3$ equations:

$$\sum_i a_i \gamma(r_i(x_j, y_j)) + b_0 + b_1 x_j + b_2 y_j = z_j \quad (j = 1, \ldots, n), \tag{3.27}$$

with

$$\sum_i a_i = 0, \quad \sum_i a_i x_i = 0, \quad \sum_i a_i y_i = 0.$$

The solution of the adjustment problem weighted by the family $\mu = \{\mu_i\}$ is the function $Z_\mu(x, y)$ of shape (3.26) such that $(8\pi a_j/\mu_j) + Z_\mu(x_j, y_j) = z_j$ $(j = 1 \ldots n)$, i.e. whose $n+3$ coefficients verify the $n+3$ equations:

$$\frac{8\pi a_j}{\mu_j} + \sum_i a_i \gamma(r_i(x_j, y_j)) + b_0 + b_1 x_j + b_2 y_j = z_j \quad (j = 1, \ldots, n), \tag{3.28}$$

with

$$\sum_i a_i = 0, \quad \sum_i a_i x_i = 0, \quad \sum_i a_i y_i = 0.$$

Moreover, these equations clearly show the interpolation problem as the limit of the adjustment problem when weights μ_j become arbitrarily large.

A surface $Z(x, y)$ of the shape (3.26) has been called a thin plate spline by reason of the physical interpretation of the integral $K(Z)$, roughly proportional to the energy of bending of a thin plate of equation $z = Z(x, y)$.

Geometric significance of the integral K(Z)

Geometrically, the integral $K(Z)$ represents the 'global' curvature of the surface $Z(x, y)$. Indeed a Z function, supposed twice differentiable at

(x, y), is approximated at the points $(x+r, y+s)$ near (x, y) by its Taylor polynomial of degree 2:

$$Z(x+r, y+s) \cong P_{x,y}(r, s) + \frac{1}{2}\left(Z''_{xx}r^2 + Z''_{xy}rs + Z''_{yx}rs + Z''_{yy}s^2\right), \quad (3.29)$$

where $P_{x,y}(r, s) = Z(x, y) + Z'_x r + Z'_y s$ is the tangent plane in (x, y).

One sees that the surface is all the more distant from its tangent plane (and so all the more concave), as $Z''_{xx}, Z''_{xy} = Z''_{yx}, Z''_{yy}$ are more distant from 0, or as $Z''^2_{xx}, Z''^2_{xy}, Z''^2_{yy}$ are larger; so the quantity $Z''^2_{xx} + 2Z''^2_{xy} + Z''^2_{yy}$ measures the curvature of the Z surface in every point.

REMARK

The quantity $Z''^2_{xx} + 2Z''^2_{xy} + Z''^2_{yy}$ has the advantage over similar measures such as $|Z''_{xx}| + |Z''_{xy}| + |Z''_{yy}|$ or $\sup(|Z''_{xx}|, |Z''_{xy}|, |Z''_{yy}|)$ of being a characteristic of the surface, invariant by a rotation of the orthonormal reference frame, i.e. by change of variable $(x, y) = (X \cos a - Y \sin a, X \sin a + Y \cos a)$. For the U parametrization of the Z surface defined by $U(X, Y) = Z(X \cos a - Y \sin a, X \sin a + Y \cos a) = Z(x, y)$, one has:

$U''_{XX}(X,Y) =$
$Z''_{xx}(x, y) \cos^2 a + 2Z''_{xy}(x, y) \sin a \cos a + Z''_{yy}(x, y) \sin^2 a$ (3.30)

$U''_{XY}(X,Y) = -Z''_{xx}(x, y) \sin a \cos a + Z''_{xy}(x, y)(\cos^2 a - \sin^2 a)$
$+ Z''_{yy}(x, y) \sin a \cos a$

$U''_{YY}(X,Y) =$
$Z''_{xx}(x, y) \sin^2 a - 2Z''_{xy}(x, y) \sin a \cos a + Z''_{yy}(x, y) \cos^2 a.$

Writing: $\sin 2A = 2Z''_{xy}/C$, $\cos 2A = (Z''_{xx} - Z''_{yy})/C$, where

$$C = \sqrt{(Z''_{xx} - Z''_{yy})^2 + 4Z''^2_{xy}},$$

one gets the expressions:

$$U''_{XX} = \frac{(Z''_{xx} + Z''_{yy})}{2} + \frac{C}{2} \cos 2(a - A)$$

$$U''_{XY} = -\frac{C}{2} \sin 2(a - A)$$

$$U''_{YY} = \frac{(Z''_{xx} + Z''_{yy})}{2} - \frac{C}{2} \cos 2(a - A) \quad (3.31)$$

from which it is easy to verify that

$$U''^2_{XX} + 2U''^2_{XY} + U''^2_{YY} = Z''^2_{xx} + 2Z''^2_{xy} + Z''^2_{yy}.$$

Justification of the approximation of the topographic surface by a thin plate spline surface

One understands now that the condition of minimality of the integral $K(Z)$ means that between sample points the 'global' curvature of the surface is as weak as possible, or else that variations of the tangent plane, and therefore of the slope, are as weak as possible. So a thin plate spline surface adjusted on contour lines presents a slope as regular as possible between contour lines, which is compatible with the definition of the topographic surface.

We should add that the maximal regularity of the thin plate spline surface applies at a time in all directions, and in particular in the horizontal direction, which notably has the effect of 'straightening' the surface exaggeratedly on crests and in thalwegs; the adjusted surface may therefore present artefacts, and then describes the topographic surface only roughly.

Implementing the approximation by a thin plate spline surface

STEP 1. CONSTITUTION OF AN APPROPRIATE SAMPLE OF POINTS

To use the method, it is necessary to first select on contour lines a sample of points, while remembering that a segment of curve $[p_{ij}, p_{i+i,j}]$ must influence the surface by all its points, and not only by its extremities. By taking as a sample only the initially given points p_{ij}, one sees that all segments would have the same influence, whereas it seems natural that a segment influences according to its length. Therefore one should cut all contours into small elementary 'segments' of similar length, in order to assure an influence proportional to the length.

STEP 2. CHOICE OF WEIGHTS IN THE CASE OF THE THIN PLATE SPLINE SURFACE OF ADJUSTMENT

A thin plate spline surface of adjustment $Z(x, y)$ by definition minimizes the quantity:

$$E(Z) = K(Z) + \sum_i \mu_i (Z(x_i, y_i) - z_i)^2,$$

preferably written

$$E_P(Z) = K(Z) + P \sum_i r_i (Z(x_i, y_i) - z_i)^2,$$

where

$$P = \sum_i \mu_i \text{ and } r_i = \frac{\mu_i}{P};$$

r_i is the relative weight of the point (x_i, y_i, z_i) and the P weight fixes the importance of the adjustment criterion

$$e(Z) = \sum_i r_i (Z(x_i, y_i) - z_i)^2$$

in relation to the curvature criterion $K(Z)$.

It is then necessary to choose the weight P and weights r_i.

(a) Let us first examine the effect of the weight P.

To $P > 0$, corresponds a unique thin plate spline surface, denoted Z_P, minimizing $E_P(Z) = K(Z) + Pe(Z)$; in the same way, to $Q > 0$ corresponds the surface Z_Q, minimizing $E_Q(Z) = K(Z) + Qe(Z)$.

Because of the minimal character of $E_P(Z_P)$ and of $E_Q(Z_Q)$, one has

$$E_P(Z_P) \leq E_P(Z_Q) \text{ and } E_Q(Z_Q) \leq E_Q(Z_P).$$

Or else:

$$K(Z_P) + Pe(Z_P) \leq K(Z_Q) + Pe(Z_Q),$$

$$K(Z_Q) + Qe(Z_Q) \leq K(Z_P) + Qe(Z_P).$$

Adding on the one hand the inequalities,

$$K(Z_P) + Pe(Z_P) \leq K(Z_Q) + Pe(Z_Q).$$

$$-K(Z_P) - Qe(Z_P) \leq -K(Z_Q) - Qe(Z_Q),$$

and on the other hand the inequalities,

$$-QK(Z_Q) - PQe(Z_Q) \leq -QK(Z_P) - PQe(Z_P),$$

$$PK(Z_Q) + PQe(Z_Q) \leq PK(Z_P) + PQe(Z_P),$$

one gets the inequalities,

$$(P - Q)e(Z_P) \leq (P - Q)e(Z_Q)$$

and

$$(P - Q)K(Z_Q) \leq (P - Q)K(Z_P), \tag{3.32}$$

showing that if $P > Q$, then $K(Z_P) \geq K(Z_Q)$ and $e(Z_P) \leq e(Z_Q)$; it means that the surface Z_Q is less bent than Z_P, or 'smoother', but not so well adjusted to the sample. In particular the interpolation thin plate spline Z_∞, limit of Z_P when P becomes arbitrarily large, is perfectly adjusted to the sample, but is less smooth than any thin plate spline of adjustment Z_Q.

(b) We suppose the P weight henceforth fixed, and we consider the weights r_i.

The simplest choice is of course to give the same weight $r_i = 1/n$ to all points.

This choice is not satisfactory because the distribution of points (x_i, y_i) can be heterogeneous because of the irregular horizontal distribution of contours, and of variations of the sampling interval along contours. Uniform weights would make the adjustment surface $Z(x, y)$ deflect toward portions of land finely sampled, which is unacceptable because Z has no reason to be affected by the sampling interval.

To avoid this inconvenience, it is therefore necessary that weights r_i are small where the sample is dense, and larger where the sample is sparse. In other words a weight r_i must be weak when the point (x_i, y_i) is representative of a small area A_i, and large when the point is representative of a large one, which brings about that r_i is an increasing function $f(|A_i|)$ of the area $|A_i|$. Besides, this function can only be linear, because if one unifies two contiguous parcels, A_i, A_j in one $A_i \cup A_j$, then the weight $f(|A_i \cup A_j|) = f(|A_i| + |A_j|)$ of the union must be the sum of weights:

$$f(|A_i| + |A_j|) = r_i + r_j = f(|A_i|) + f(|A_j|). \tag{3.33}$$

Thus, $r_i = k|A_i|$, and the condition $\sum r_i = 1$ imposes

$$k = \frac{1}{\sum |A_i|}.$$

It remains to specify what is a parcel A_i represented by the point (x_i, y_i, z_i); it is natural to define it as the set of points (x, y) that are closer to (x_i, y_i) than to the other (x_j, y_j) points; one knows that this set is a polygon V_i and that the family $\{V_i; i = 1, \ldots, n\}$ forms the Voronoï diagram associated with points (x_i, y_i).

Finally, a coherent system of weight can be defined by:

$$r_i = \frac{|V_i|}{\sum_j |V_j|}, \tag{3.34}$$

where $|V_i|$ designates the area of the Voronoï polygon V_i.

STEP 3. RESOLUTION OF THE LINEAR SYSTEM

The following step consists in solving the linear system of $n+3$ equations giving coefficients $a_1, \ldots, a_n, b_0, b_1, b_2$.

Denoting $c_{ij} = r_i(x_j, y_j)^2 \ln r_i(x_j, y_j)$ this system becomes in matrix form:

$$\begin{pmatrix} 8\pi/\mu_1 & c_{21} & \cdots & c_{n1} & 1 & x_1 & y_1 \\ c_{12} & 8\pi/\mu_2 & \cdots & c_{n2} & 1 & x_2 & y_2 \\ \cdots & \cdots & \cdots & \cdots & \cdots & \cdots & \cdots \\ c_{1n} & c_{2n} & \cdots & 8\pi/\mu_n & 1 & x_n & y_n \\ 1 & 1 & \cdots & 1 & 0 & 0 & 0 \\ x_1 & x_2 & \cdots & x_n & 0 & 0 & 0 \\ y_1 & y_2 & \cdots & y_n & 0 & 0 & 0 \end{pmatrix} \begin{pmatrix} a_1 \\ a_2 \\ \cdots \\ a_n \\ b_0 \\ b_1 \\ b_2 \end{pmatrix} = \begin{pmatrix} z_1 \\ z_2 \\ \cdots \\ z_n \\ 0 \\ 0 \\ 0 \end{pmatrix}$$

which can be written as

$$\begin{pmatrix} C & F \\ F^T & 0 \end{pmatrix} \begin{pmatrix} a \\ b \end{pmatrix} = \begin{pmatrix} Z \\ 0 \end{pmatrix}. \tag{3.35}$$

Note that,

$$r_i(x_j, y_j) = \sqrt{(x_j - x_i)^2 + (y_j - y_i)^2} = r_j(x_i, y_i),$$

therefore $c_{ji} = c_{ij}$; the matrix C is therefore symmetrical.

Note that neither the matrix

$$\begin{pmatrix} C & F \\ F^T & 0 \end{pmatrix}$$

nor its opposite, are positive; for example, if $a = (1, 0, 0, \ldots, 0)$ and $b = (b_0, 0, 0)$, one gets:

$$(a^T \ b^T) \begin{pmatrix} C & F \\ F^T & 0 \end{pmatrix} \begin{pmatrix} a \\ b \end{pmatrix} = \frac{8\pi}{\mu_1} + 2b_0$$

which can be ≥ 0 or ≤ 0 according to the value of b_0.

C. Carasso in Baranger (1991) suggests solving the system (3.35) as follows. One factorizes F as $F = QR$ where Q is an (n, n) orthogonal matrix, and R an $(n, 3)$ matrix in the shape

$$R = \begin{pmatrix} U \\ 0 \end{pmatrix},$$

with U a $(3, 3)$ triangular superior matrix. F is of rank 3, since there are at least three non-aligned points, therefore R is of rank 3, and thus U is invertible. So, one can write (here I is the $(3, 3)$ identity matrix):

$$\begin{pmatrix} C & F \\ F^T & 0 \end{pmatrix} = \begin{pmatrix} Q & 0 \\ 0 & I \end{pmatrix} \begin{pmatrix} Q^TCQ & R \\ R^T & 0 \end{pmatrix} \begin{pmatrix} Q^T & 0 \\ 0 & I \end{pmatrix} \quad (3.36)$$

so that the system (3.35) is equivalent to the system

$$\begin{pmatrix} Q^TCQ & R \\ R^T & 0 \end{pmatrix} \begin{pmatrix} d \\ b \end{pmatrix} = \begin{pmatrix} Q^TZ \\ 0 \end{pmatrix}, \quad a = Qd. \quad (3.37)$$

Writing by blocks, $Q = (Q_1 \ Q_2)$, $Q_1(n, 3)$ matrix, $Q_2(n, n-3)$ matrix, and

$$d = \begin{pmatrix} d_1 \\ d_2 \end{pmatrix}, \quad d_1(3) \text{ vector, } d_2(n-3) \text{ vector,}$$

this system is written:

$$\begin{pmatrix} Q_1^TCQ_1 & Q_1^TCQ_2 & U \\ Q_2^TCQ_1 & Q_2^TCQ_2 & 0 \\ U^T & 0 & 0 \end{pmatrix} \begin{pmatrix} d_1 \\ d_2 \\ b \end{pmatrix} = \begin{pmatrix} Q_1^TZ \\ Q_2^TZ \\ 0 \end{pmatrix}, \quad a = Q_1d_1 + Q_2d_2; \quad (3.38)$$

one deduces $U^Td_1 = 0$, thus $d_1 = 0$; from the equation $Q_2^TCQ_2d_2 = Q_2^TZ$, one deduces d_2, from where $a = Q_2d_2$; from the equation $Q_1^TCQ_2d_2 + Ub = Q_1^TZ$, or $Ub = Q_1^T(Z - Ca)$, one finally extracts b.

It is necessary to notice that the C matrix is full, therefore the matrix $Q_2^TCQ_2$ also; one will see that to the contrary the 'elastic' grid method drives to a sparse system, allowing larger size systems to be processed with the same memory resources.

STEP 4. CONSTITUTION OF THE REGULAR GRID

This immediate step consists in calculating, by the formula (3.26), the altitudes $z_{cl} = Z(x_c, y_l)$ at the grid nodes.

3.5.2.4 Approximation of the topographic surface by an elastic grid surface (de Masson d'Autume, 1978)

Definition

An elastic grid surface can be introduced as a discrete approximation of a thin plate spline surface of adjustment $Z(x, y)$, defined in §3.5.2.3; one knows that $Z(x, y)$ minimizes a quantity:

$$E(Z) = K(Z) + \sum_i \mu_i \big(Z(x_i, y_i) - z_i\big)^2. \tag{3.39}$$

As our final objective is not the $Z(x, y)$ surface, but only the set of its values $z_{cl} = Z(x_c, y_l)$ at the nodes of a regular grid $\{x_c = ch; c = 1, \ldots, N\} \times \{y_l = lh; l = 1, \ldots, M\}$, one can try to set a problem whose unknown is the set $\{z_{cl}\}$ directly. Thus we shall replace the $E(Z)$ quantity by an approached quantity that is a function of the z_{cl} only.

(1) On the one hand one has to replace the integral

$$K(Z) = \int Z''^2_{xx} + 2Z''^2_{xy} + Z''^2_{yy}\, dx\,dy.$$

One first approaches it by Riemann sum:

$$K(Z) \cong h^2 \sum_{c,l} Z''_{xx}(x_c, y_l)^2 + 2Z''_{xy}(x_c, y_l)^2 + Z''_{yy}(x_c, y_l)^2, \tag{3.40}$$

where $c = 1 : N, l = 1 : M$.

One can then write the following eight Taylor formulae at point (x_c, y_l):

$$z_{c+1,l} - z_{cl} = Z'_x h + Z''_{xx} \frac{h^2}{2} + Z'''_{xxx} \frac{h^3}{6} + o(h^3)$$

$$z_{c-1,l} - z_{cl} = -Z'_x h + Z''_{xx} \frac{h^2}{2} - Z'''_{xxx} \frac{h^3}{6} + o(h^3)$$

$$z_{c,l+1} - z_{cl} = Z'_y h + Z''_{yy} \frac{h^2}{2} + Z'''_{yyy} \frac{h^3}{6} + o(h^3)$$

$$z_{c,l-1} - z_{cl} = -Z'_y h + Z''_{yy} \frac{h^2}{2} - Z'''_{yyy} \frac{h^3}{6} + o(h^3)$$

$$z_{c+1,l+1} - z_{cl} = Z'_x h + Z'_y h + (Z''_{xx} + 2Z''_{xy} + Z''_{yy}) \frac{h^2}{2} +$$
$$(Z'''_{xxx} + 3Z'''_{xxy} + 3Z'''_{xyy} + Z'''_{yyy}) \frac{h^3}{6} + o(h^3)$$

$$-z_{c+1,l-1} + z_{cl} = -Z'_x h + Z'_y h + (-Z''_{xx} + 2Z''_{xy} + Z''_{yy}) \frac{h^2}{2} +$$
$$(-Z'''_{xxx} + 3Z'''_{xxy} - 3Z'''_{xyy} + Z'''_{yyy}) \frac{h^3}{6} + o(h^3)$$

$$-z_{c-1,l+1} + z_{cl} = Z'_x h + Z'_y h + (-Z''_{xx} + 2Z''_{xy} - Z''_{yy}) \frac{h^2}{2} +$$

$$(Z'''_{xxx} - 3Z'''_{xxy} + 3Z'''_{xyy} - Z'''_{yyy})\frac{h^3}{6} + o(h^3)$$

$$z_{c-1,l-1} - z_{cl} = -Z'_x h - Z'_y h + (Z''_{xx} + 2Z''_{xy} + Z''_{yy})\frac{h^2}{2} +$$

$$(-Z'''_{xxx} - 3Z'''_{xxy} - 3Z'''_{xyy} - Z'''_{yyy})\frac{h^3}{6} + o(h^3); \quad (3.41)$$

adding the first and second formulae, then the third and fourth, and finally the last four, one deduces:

$$Z''_{xx}(x_c, y_l) = \frac{(z_{c-1,l} - 2z_{cl} + z_{c+1,l})}{h^2} + o(h)$$

$$Z''_{yy}(x_c, y_l) = \frac{(z_{c,l-1} - 2z_{cl} + z_{c,l+1})}{h^2} + o(h)$$

$$Z''_{xy}(x_c, y_l) = \frac{(z_{c-1,l-1} - z_{c-1,l+1} - z_{c+1,l-1} + z_{c+1,l+1})}{4h^2} + o(h). \quad (3.42)$$

These values permit us to approximate the integral $K(Z)$ by the quantity function of the vector $\mathbf{z} = \{z_{cl}\}$ of R^{MN}:

$$K_h(\mathbf{z}) =$$

$$\frac{1}{h^2}\left(\sum_{c=2}^{N-1}\sum_{l=1}^{M}(z_{c-1,j} - 2z_{cl} + z_{c+1,l})^2 + \sum_{c=1}^{N}\sum_{l=2}^{M-1}(z_{c,l-1} - 2z_{cl} + z_{c,l+1})^2\right.$$

$$\left.\sum_{c=2}^{N-1}\sum_{l=2}^{M-1}(z_{c-1,l-1} - z_{c-1,l+1} - z_{c+1,l-1} + z_{c+1,l+1})^2/8\right). \quad (3.43)$$

(2) On the other hand one has to replace in

$$e(Z) = \sum_i \mu_i(Z(x_i, y_i) - z_i)^2$$

every $Z(x_i, y_i)$ by a function of the vector $\mathbf{z}$.

One will adopt a representation by a piecewise bilinear or bicubic function

$$Z(x, y) = \sum_{c,l} z_{cl}\, V\!\left(\frac{x}{h} - c\right) V\!\left(\frac{y}{h} - l\right)$$

with, in the bilinear case, $V(t) = Q(t) = \sup(1 - |t|, 0)$ and in the bicubic case:

$$V(t) = U(t) = \begin{cases} \frac{3}{2}|t|^3 - \frac{5}{2}|t|^2 + 1 & \text{if } |t| \leq 1 \\ -\frac{1}{2}|t|^3 + \frac{5}{2}|t|^2 - 4|t| + 2 & \text{if } 1 \leq |t| \leq 2 \\ 0 & \text{if } |t| \geq 2 \end{cases} \quad (3.44)$$

$e(Z)$ is then replaced by

$$e_h(Z) = \sum_i \mu_i \left(\sum_{c,l} z_{cl} V\left(\frac{x_i}{h} - c\right) V\left(\frac{y_i}{h} - l\right) - z_i \right)^2.$$

Finally the $E(Z)$ quantity is replaced by the quantity function of

$$\mathbf{z} = \{z_{cl}\}, \quad E_h(\mathbf{z}) = K_h(\mathbf{z}) + e_h(\mathbf{z}).$$

One defines the elastic grid surface associated with the sample $\{(x_i, y_i, z_i); i = 1, \ldots, n\}$ as the surface

$$Z(x, y) = \sum_{c,l} z_{cl} V\left(\frac{x}{h} - c\right) V\left(\frac{y}{h} - l\right)$$

defined by the vector $\mathbf{z} = \{z_{cl}\}$ that minimizes the quantity $E_h(\mathbf{z})$.

Existence and unicity of the solution for the minimization problem of $E_h(\mathbf{z})$

The quantity $E_h(\mathbf{z})$ is a quadratic function of $\mathbf{z} = \{z_{cl}\}$, which therefore can be minimized explicitly.

In a precise way, one introduces the vectors $C_{cl}, D_{cl}, F_{cl}, B_i$ such as

$$z_{c-1,l} - 2z_{cl} + z_{c+1,l} = C_{cl}^T \mathbf{z}$$

$$z_{c,l+1} - 2z_{cl} + z_{c,l+1} = D_{cl}^T \mathbf{z}$$

$$z_{c-1,l-1} - z_{c-1,l+1} - z_{c+1,l-1} + z_{c+1,l+1} = F_{cl}^T \mathbf{z}$$

$$\sum_{c,l} z_{cl} V\left(\frac{x_i}{h} - c\right) V\left(\frac{y_i}{h} - l\right) = B_i^T \mathbf{z}, \qquad (3.45)$$

for example,

$$C_{cl}^T = (0 \quad \ldots \quad 0 \quad \underset{c-1,l}{1} \quad \underset{cl}{-2} \quad \underset{c+1,l}{1} \quad 0 \quad \ldots \quad 0).$$

One then sees that:

$$K_h(\mathbf{z}) = \frac{1}{h^2} \left(\sum_{cl} \mathbf{z}^T C_{cl} C_{cl}^T \mathbf{z} + \sum_{cl} \mathbf{z}^T D_{cl} D_{cl}^T \mathbf{z} + \frac{1}{8} \sum_{cl} \mathbf{z}^T F_{cl} F_{cl}^T \mathbf{z} \right)$$

$$e_h(\mathbf{z}) = \sum_i \mu_i (\mathbf{z}^T B_i - z_i)(B_i^T \mathbf{z} - z_i) =$$

$$\mathbf{z}^T \left(\sum_i \mu_i B_i B_i^T \right) \mathbf{z} - 2 \left(\sum_i \mu_i z_i B_i \right)^T \mathbf{z} + \sum_i \mu_i z_i^2, \qquad (3.46)$$

so that, putting:

$$A = \frac{1}{h^2}\left(\sum_{cl} C_{cl}C_{cl}^T + \sum_{cl} D_{cl}D_{cl}^T + \frac{1}{8}\sum_{cl} F_{cl}F_{cl}^T\right) + \sum_i \mu_i B_i B_i^T$$

$$b = \sum_i \mu_i z_i B_i \quad c = \sum_i \mu_i z_i^2 \tag{3.47}$$

one can write $E_h(\mathbf{z})$ as:

$$E_h(\mathbf{z}) = \mathbf{z}^T A \mathbf{z} - 2b^T \mathbf{z} + c.$$

The matrix A is symmetrical. It is positive because

$$\mathbf{z}^T A \mathbf{z} = \frac{1}{h^2}\left(\sum_{cl}(C_{cl}^T \mathbf{z})^2 + \sum_{cl}(D_{cl}^T \mathbf{z})^2 + \frac{1}{8}\sum_{cl}(F_{cl}^T \mathbf{z})^2\right)$$

$$+ \sum_i \mu_i (B_i^T \mathbf{z})^2 \geq 0$$

whatever the value of $\mathbf{z}$.

One intends to show that the matrix A is positive definite, which means that $\mathbf{z}^T A \mathbf{z}$ can be 0 only if $\mathbf{z} = 0$; this property will guarantee the existence and the unicity of the solution of the problem.

Let us suppose therefore that $\mathbf{z}^T A \mathbf{z}$ for a certain $\mathbf{z}$.

Then $C_{cl}^T \mathbf{z} = 0$ for any c, l, which implies $z_{cl} = z_{1l} + (c - 1)(z_{2l} - z_{1l})$

In the same way $D_{cl}^T \mathbf{z} = 0$ for any c, l, which implies $z_{cl} = z_{c1} + (l - 1)(z_{c2} - z_{c1})$.

One deduces then:

$$\begin{aligned} z_{cl} &= z_{11} + (c-1)(z_{21} - z_{11}) + (l-1)(z_{12} - z_{11}) + \\ &\quad (c-1)(l-1)(z_{22} - z_{12} - z_{21} - z_{11}) \\ &= a_0 + a_1 c + a_2 l + a_3 cl \end{aligned} \tag{3.48}$$

for any c, l;

from where

$$F_{cl}^T \mathbf{z} = z_{c+1,l+1} - z_{c+1,l-1} - z_{c-1,l+1} + z_{c-1,l-1} = 4a_3.$$

The hypothesis $\mathbf{z}^T A \mathbf{z} = 0$ also implies $F_{cl}^T \mathbf{z}$ for any c, l, from where $a_3 = 0$.

Finally: $z_{cl} = a_0 + a_1 c + a_2 l$ for any c, l.

248 Grégoire Maillet et al.

One has then:

$$\begin{aligned}
B_i^T z &= \sum_{cl} (a_0 + a_1 c + a_2 l)\, V\left(\frac{x_i}{h} - c\right) V\left(\frac{y_i}{h} - l\right) \\
&= a_0 \sum_c V\left(\frac{x_i}{h} - c\right) \sum_l V\left(\frac{y_i}{h} - l\right) \\
&\quad + a_1 \sum_c cV\left(\frac{x_i}{h} - c\right) \sum_l V\left(\frac{y_i}{h} - l\right) \\
&\quad + a_2 \sum_c V\left(\frac{x_i}{h} - c\right) \sum_l lV\left(\frac{y_i}{h} - l\right).
\end{aligned}$$

However, one can show that the V function, when it designates the Q linear interpolation function as well as the U cubic interpolation function, verifies the identities:

$$\sum_c V(x - c) = 1 \text{ and } \sum_c cV(x - c) = x \text{ for any } x;$$

from where:

$$B_i^T z = a_0 + a_1 \frac{x_i}{h} + a_2 \frac{y_i}{h}.$$

But the hypothesis $z^T A z = 0$ implies $B_i^T z = a_0 + a_1(x_i/h) + a_2(y_i/h) = 0$ for any i; now one can suppose that the sample counts at least three (x_j, y_j), (x_k, y_k), (x_p, y_p) non-aligned points; these three points verify the system

$$\left\{ a_0 + a_1 \frac{x_j}{h} + a_2 \frac{y_j}{h} = 0,\ a_0 + a_1 \frac{x_k}{h} + a_2 \frac{y_k}{h} = 0,\ a_0 + a_1 \frac{x_p}{h} + a_2 \frac{y_p}{h} = 0 \right\}$$

of non-null determinant $(x_p - x_j)(y_k - y_j) - (x_k - x_j)(y_p - y_j)$, which is possible only if $a_0 = a_1 = a_2 = 0$. Therefore $z = 0$; one has thus shown that A is positive definite.

A is then invertible, and one can write therefore:

$$E_h(z) = (z - A^{-1}b)^T A(z - A^{-1}b) + c - b^T A^{-1} b \geq c - b^T A^{-1} b,$$

which shows that $E_h(z)$ reaches its minimum $c - b^T A^{-1} b$ if and only if $z = A^{-1} b$. In other words the problem of minimization of $E_h(z)$ admits a solution z and only one given by the equation $Az = b$.

Structure of the A matrix

One has: $A = A_1 + A_2$ where

$$A_1 = \frac{1}{h^2}\left(\sum_{cl} C_{cl}C_{cl}^T + \sum_{cl} D_{cl}D_{cl}^T + \frac{1}{8}\sum_{cl} F_{cl}F_{cl}^T\right), \quad A_2 = \sum_i \mu_i B_i B_i^T,$$

A, A_1, A_2 are (MN, MN) matrixes (M: number of lines, N: number of columns of the grid).

If we expand A, we get:

$$A_1 = \frac{1}{h^2}\begin{pmatrix} I+X+\frac{Y}{8} & -2I & I-\frac{Y}{8} \\ -2I & 5I+X+\frac{Y}{8} & -4I & I-\frac{Y}{8} \\ I-\frac{Y}{8} & -4I & 6I+X+\frac{Y}{4} & -4I & I-\frac{Y}{8} \\ & \cdots & \cdots & \cdots & \cdots & \cdots \\ & & I-\frac{Y}{8} & \cdots & 6I+X+\frac{Y}{4} & -4I & I-\frac{Y}{8} \\ & & & I-\frac{Y}{8} & -4I & 5I+X+\frac{Y}{8} & -2I \\ & & & & I-\frac{Y}{8} & -2I & I+X+\frac{Y}{8} \end{pmatrix}$$

where I is the (N, N) identity matrix, X and Y are (N, N) matrixes:

$$X = \begin{pmatrix} 1 & -2 & 1 \\ -2 & 5 & -4 & 1 \\ 1 & -4 & 6 & -4 & 1 \\ \cdots & \cdots & \cdots & \cdots & \cdots \\ & & 1 & -4 & 6 & -4 & 1 \\ & & & 1 & -4 & 5 & -2 \\ & & & & 1 & -2 & 1 \end{pmatrix}$$

$$Y = \begin{pmatrix} 1 & 0 & -1 \\ 0 & 1 & 0 & -1 \\ -1 & 0 & 2 & 0 & -1 \\ \cdots & \cdots & \cdots & \cdots & \cdots \\ & & -1 & 0 & 2 & 0 & -1 \\ & & & -1 & 0 & 1 & 0 \\ & & & & -1 & 0 & 1 \end{pmatrix}.$$

Expanding A_2 in the case of bilinear model one gets:

$$A_2 = \begin{pmatrix} X & X \\ X & X & X \\ \cdots & \cdots & \cdots \\ & & X & X & X \\ & & & X & X \end{pmatrix}$$

where each X is a matrix of the shape

$$X = \begin{pmatrix} \bullet & \bullet & & & \\ \bullet & \bullet & \bullet & & \\ \ldots & \ldots & \ldots & & \\ & & \bullet & \bullet & \bullet \\ & & & \bullet & \bullet \end{pmatrix}$$

A_1 and A_2 are written in the shape of (M, M) matrixes of (N, N) size blocks.

Every line of the matrix A includes a maximum of 21 non-null coefficients on a total of MN; for a grid $M = 100$, $N = 100$, there are therefore more than 99.79 per cent of zero coefficients. One sees that A is really a sparse matrix, as stated.

References

Carasso C. (1991) Lissage des données à l'aide de fonctions spline, in J. Baranger (ed.) *Analyse numérique*, Hermann, pp. 357–414.

de Masson d'Autume G. (1978) Construction du modèle numérique d'une surface par approximations successives. Application aux modèles numériques de terrain (M.N.T.), *Bulletin de la Société française de photogrammétrie et de télédétection*, no. 71–72, pp. 33–41.

de Masson d'Autume G. (1979) Surface modelling by means of an elastic grid, *Photogrammetria*, no. 35, pp. 65–74.

Duchon J. (1976) Interpolation des fonctions de deux variables suivant le principe de la flexion des plaques minces, *Revue Française d'Automatique, Informatique et Recherche Opérationnelle (R.A.I.R.O.) Analyse numérique*, vol. 10, no. 12, décembre, pp. 5–12.

3.5.3 Merging contour lines and 3D measurements from images

Nicolas Paparoditis

The two techniques that have been described in the previous paragraphs, for understandable pedagogic and clarity reasons, have been developed separately. Nevertheless, the DSM reconstruction from contour lines can be improved by adding 3D points processed by photogrammetric means to solve ambiguities between all possible triangulation of contour lines. In the same way, contour lines can be injected as constraint lines in the process of triangulation of all 3D points that can processed from the images. Both techniques lead to a constrained triangulation. Nevertheless, these merging processes will have the chance of succeeding if both 3D information are homogeneous in accuracy and precision and definitely describe the same surface.

3.5.4 Generating a DTM with sparse 3D points and characteristic features

Nicolas Paparoditis

A DTM can be generated manually by plotting well-chosen sparse on ground 3D points in a stereo display. The spatial localization of these points is determined by the operator and their elevation can be also manually plotted or automatically generated through the stereo matching methods we have discussed earlier. The spatial density of the points depends on the local roughness of the landscape and on the desired DTM quality. A TIN-like surface can be generated from these points with a 2D Delaunay triangulation, as shown in Figure 3.5.4.1 (a) and (b).

Instead of plotting much denser points (increasing production times) around terrain break lines to render locally the relief morphology, the operator can plot the break lines themselves and use these lines to constrain locally the triangulation. In practice, we destroy all triangles overlapping

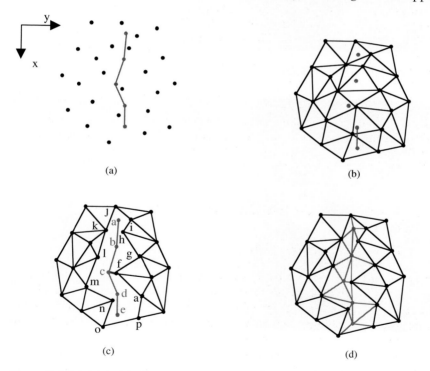

Figure 3.5.4.1 (a) in black sparse cloud of 3D points and in grey the slope beak line; (b) 2D triangulation of all points in the cloud; (c) all triangles overlapping the break line are removed and (d) the remaining space is triangulated under the constraint that the new triangles lie on the break line segments.

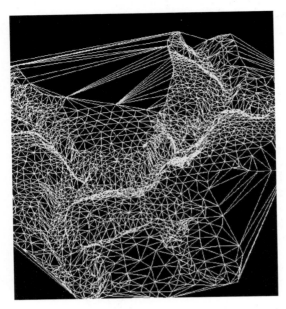

Figure 3.5.4.2 (a) Triangulation without characteristic constraint lines.

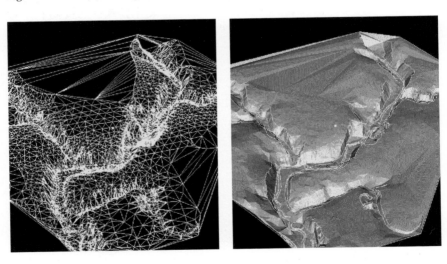

Figure 3.5.4.2 (b) Triangulation with characteristic constraint lines.

the break line and we triangulate the remaining space. For instance, in the case of an open slope break line as shown on Figure 3.5.4.1, the triangulation of this empty space can be seen as the triangulation of the polygon (c, f, g, h, k, l, m, n, o, p, f, c, d, e, d, c, b, a, b, c) in the Figure 3.5.4.1 (c) example, where c f is the smallest segment joining the break line points and the surrounding triangle points.

Figures 3.5.4.2 (a) and (b) show some triangulation results on a real set of data (points and characteristic terrain line) plotted on a digital photogrammetric workstation.

3.6 EXTRACTION OF CHARACTERISTIC LINES OF THE RELIEF

Alain Dupéret, Olivier Jamet

3.6.1 Some definitions bound to the relief

3.6.1.1 Preliminary definitions

The building of the relief results from many endogenous actions and exogenous phenomena such as lithology, climatic variations, vegetation, anthropic activities. ... The hydrographic network is a set of particularly useful morphological objects to characterize the relief. The description of this last by the characteristic elements leans on a mathematical representation that depends on:

- the type of data capture: the surface itself, the cartographic representation;
- the mode of acquisition: direct surveys, digitization, photogrammetric restitution;
- the mode of altitudes distribution: contour line, regular or irregular set of points.

In practice, it is desirable to do a separation between aspects bound to the acquisition of the ground characteristic elements and those bound to the mathematical representation of the surface. Otherwise, notions of thalwegs and rivers are often assimilated, and this leads to confusions such as how to assimilate the relief and the process of water streaming, even if they are the most often linked; the superficial outflow most often takes place in the sense of the local or regional line of stronger slope, but the geology (by the presence of hard or soft rock) and the fracturation can also intervene.

First, a simple topological mathematical presentation will be given for the main remarkable elements of the relief. Then, several operative modes will be presented on the basis of some important studies.

3.6.1.2 Definitions of the main characteristic elements of the relief

The definitions that follow are based upon a quite simplistic modelization of the reality: we will suppose that the topographic surface is correctly described by a function $z = H(x, y)$, continuous and twice derivable. Obviously it is only an approximation, as the topography presents many similarities with fractal structures, and presents discontinuities at every scale of observation. Nevertheless, this approach allows for much simpler definitions of the characteristic elements of the relief.

The main remarkable points

THE SUMMIT

The summit is the high point isolated representative of the local maximum of the H function. In practice, one generally uses the following (incomplete) characterization:

$$(X_0, Y_0) \text{ is a summit if } \begin{cases} H'_x(X_0, Y_0) = 0 \\ H'_y(X_0, Y_0) = 0 \\ H''_{xx}(X_0, Y_0) < 0 \text{ and } H''_{yy}(X_0, Y_0) < 0 \end{cases}$$

THE BASIN

The basin is the low point isolated representing the local minimum of the H function. In practice:

$$(X_0, Y_0) \text{ is a basin if } \begin{cases} H'_x(X_0, Y_0) = 0 \\ H'_y(X_0, Y_0) = 0 \\ H''_{xx}(X_0, Y_0) > 0 \text{ and } H''_{yy}(X_0, Y_0) > 0 \end{cases},$$

which means therefore that radii of curvatures in the principle planes are negative.

THE PASS

The pass is an origin point of at least two divergent water streams. The notion of pass evokes a point as much as a surface. From the macroscopic point of view, it is a region in which two divergent valleys separating two summits meet that is, mathematically, a saddle point. In practice:

$$(X_0, Y_0) \text{ is a pass if } \begin{cases} H'_x(X_0, Y_0) = 0 \\ H'_y(X_0, Y_0) = 0 \\ H''_{xx}(X_0, Y_0) \cdot H''_{yy}(X_0, Y_0) < 0 \end{cases}.$$

This situation implies the existence of such a point, that any infinitesimal circle around it shows at least a double slope alternance. The water stream may, from a minute displacement, go in two different directions.

The detection of passes according to these criteria often produces in practice many parasitic passes, of weak amplitude, close to an extremum of H. Most often, a minimal modification of altitudes in the DTM makes such artefacts disappear. The replacement of a square mesh by a triangular mesh would also annul a certain number of them. This purification of the model is a prerequisite for a correct topographic modelling and an easier extraction of the characteristic lines.

The main characteristic lines

From a global point of view, a characteristic line is a line of slope assuring the sharing of the outflow of the streaming waters. The definitions of these lines often present two variants, depending on whether they are understood according to the real geomorphology, or according to the algorithms used to detect them. For example, with the geomorphological meaning, a thalweg is generally not formed at the beginning of the watershed of a torrent, the opposite of what the determination algorithms may show, as they may define a 'theoretical' thalweg that may climb up to the pass.

THE MAIN THALWEG

The thalweg can be determined by the journey of a stream between the pass that is maybe at its origin, to its end naturally in the basin that it is going to find on its journey, while following the line of stronger downward slope.

THE MAIN CREST

The localization of every crest, in the same way, is achieved as being the journey followed by an anti-streaming (i.e. reversed streaming, assimilated to water that would go up exclusively along the line of stronger slope) from a pass to a summit, while following the line of stronger ascending slope.

CONCLUSION ON THE MAIN CHARACTERISTIC LINES

Crests and main thalwegs constitute the main oro-hydrographic system that relies completely on the preliminary determination of the passes. Special attention must be paid to low and/or flooded zones such as marshes, lakes and large flat valleys, in particular when automatic extraction is concerned.

All goes well if a tracing of a crest started in a convex zone is constituted in a convex zone up to the junction with another crest or up to a summit, but problems arise if there is an extension in a concave zone. A symmetric remark can be made for a thalweg. This justifies the definition and the taking into account of new characteristic elements of the relief.

3.6.1.3 Definitions of remarkable elements of the relief

The contour line

Contour lines are the plane curves of the function $Z = H(x, y)$. Associated to the selected and regularly spaced altitudes of an interval called equidistance of curves, they are often used to provide the cartographic representation of the relief, often with a set of edition spot heights distributed on remarkable planimetric points. Beyond this cartographic setting, the horizontals can become characteristic elements of the relief as soon as it contains extents of water. Contours of ponds and the foreshores of seas are typically lines that present a remarkable slope discontinuity in the altitude function for large topographic areas.

The curve $\Gamma n = 0$

The horizontal curvature of contour lines can also become the object of an analysis. The convention, of course, foresees that an observer who constantly follows a contour line in the direct sense turns on his right in convex ground (left in concave). Numerous difficulties, notably those bound to the parasitic passes, can be avoided by considering the limit between the concave and convex zones of the ground. While calling Γn the horizontal curvature of contour lines, the curvature $\Gamma n = 0$ represents the limit we look for. In principle, the pass is a part of this particular curve.

The secondary remarkable points

Points whose definitions follow correspond to places of null curvature.

THE THALWEG ORIGIN

To palliate annoyances mentioned previously concerning the main characteristic lines, the high points of the curve $\Gamma n = 0$ will be taken as thalweg origins.

THE END OF CREST

The crest ends will be defined as the low points of the curve $\Gamma n = 0$.

Extraction of characteristic lines of the relief   257

REMARKS ON THE SECONDARY REMARKABLE POINTS

Other approaches than the one presented are possible, as the one that would consist in taking all points for which the slope p is maximum; ends of crest are indicated when $\Gamma n > 0$, beginnings of thalweg by $\Gamma n < 0$.

The secondary characteristic lines

The secondary characteristic lines are obtained by drawing downwards from a secondary remarkable point, the lines of slope downwards or upwards.

THE SECONDARY THALWEG

The secondary thalwegs are drawn in a concave zone $\Gamma n < 0$, from the secondary thalwegs' origins, while going down along the line of stronger slope up to the confluence with another thalweg.

THE SECONDARY CREST

The secondary crests are drawn in concave zones $\Gamma n > 0$, from the secondary crest ends, while going up the long of the larger slope line up to the junction with a main line.

REMARKS ON THE SECONDARY CHARACTERISTIC LINES

The reliable extraction of the characteristic elements has therefore a particular importance, whose automation appears desirable, especially because the manual extraction seems delicate. Indeed, in opposition to the acquisition of level contour lines for which the z device is blocked at the time of the acquisition, or to the seizure of nodes of a net for which the planimetric position is imposed, the direct photogrammetric acquisition does not show any systematism. The positioning freedom is large, extremely correlated to the altimetric precision of pointing and intervenes in zones that require a significant effort to be correctly described: vegetation in thalweg bottoms, flat crests, etc.

3.6.1.4 Notions of drainage network and watershed

A drainage network is defined (Deffontaines and Chorowicz, 1991) as being composed of 'the set of the topographic surfaces situated below all neighbouring high points, generally flowing out according to a direction. These surfaces can contain water in temporary or permanent manner'. It includes:

- thalwegs: valley bottoms, narrow or large, with water or dry;
- closed endorheic or exorheic depressions such as marshes, sink-holes, lakes.

To do some automatic extractions following this definition, several families of methods are used on DTM, either by dynamic analysis of the streaming, or by local analysis of the surface curvature.

In hydrology, the true catchment area relative to a point is defined as the totality of the topographic surface drained by this point; it is therefore a geographical entity on which the ensemble of water enters, due to the rain, and shares and organizes itself to lead to a unique outlet in the exit of a catchment area. As for the topographic catchment area relative to a point, this is the location of the points of the topographic surface whose line of larger slope passes by this point. There is not necessarily coincidence of the two types of catchment areas, for example for karstic reliefs where water that streams inside a catchment area can be found in the exit in another catchment area because of losses and re-emergences. In the same way, water that infiltrates the ground can meet the impervious layers that direct it to water-tables participating thus in other hydrologic basin balance. To finish, human activity generates obstacles or reaches that often modify the natural logic of the stream.

As the topographic, geological and pedological properties of a catchment area constitute some essential parameters for its survey, it is nesessary to perform a reliable digital modelling of the relief, in particular of the drainage network. In such a context, many digital indexes, more thematic, are used to characterize a catchment area: the drained surface, lengths of drainage, the perimeter, the distance to the outlet, the size of the drained surfaces, densities and frequency of drainage, the indication of compactness, the ratio of confluence or concentration, the Beven index, etc. [Depraetère and Moniod, 1991].

3.6.2 Extraction of thalwegs from starting points

This method, named 'structuralist approach' by the author (Riazanoff et al., 1992) is inspired by the physical model of water streaming on a relief and proposes a dynamic method of determination of crest lines. The tracing of lines is judiciously dynamic from the starting points chosen, while following the line of larger slope up to arrive either in a border of the zone, a local minimum or on an already drawn line. The algorithm proceeds by two distinct steps. The first consists in describing completely the network coming down from all main passes and following the largest slope. The second applies to correcting shortcomings produced by depressions while forcing the drainage toward the lowest pass.

The thalweg is defined as 'concave place of water convergence'. The crest is defined in a dual manner as 'convex place of anti-streaming convergence'. These two characteristic line types cross themselves in singular points, mainly passes, but also the local extrema, the high points (local maximum of a concave zone) and the low points (minimum local of a convex zone).

A constraint of initial progession is applied for the particular case of the starting point. Three different advance constraints have been used for this method, the description being made for the extraction of crests.

1   The algorithm of the stream: the constraint of advance is 'to climb according to the line of larger slope', while choosing among neighbours the one that presents the largest denivelation in relation to the displacement among the neighbouring points. The starting point is a pass. Thus the network obtained can be considered as too dense. The detection is also meaningful for zones situated at a low altitude, if compared to zones of average or high altitude.
2   The algorithm of the wanderer: from the local maxima, the constraint of advance used here is 'to go down towards points presenting a convex slope change in one of the three directions (four minus the one of origin)'. This process makes it possible to survey the main structures but is very sensitive to the least disconnection that leads to the non-consideration of possible structure situated downstream.
3   The main pass algorithm adds up the information from the two previous algorithms in order to select starting points susceptible of generating the main characteristic lines. From the passes not marked in the network obtained by the previous method, the constraint of advance is 'to climb according to the line of larger slope'. All marked passes are passes not belonging to the network generated by the climbing from the local minima on thalwegs; in fact, a pass will not be marked if it is located on the extremity of a branch of the network. The main crests are correctly marked. The principal massifs and hills are marked.

The third method, using the first two, gives best results. The outlet is searched for. Logically, the point of outlet is the lowest pass among those from where come lines of outflow arriving at this depression. To improve the method, the sense of outflow of the line from the outlet is reversed, and the outflow continues on the other side of the pass. Other approaches preserved the principle to apply a method of streaming according to the line of larger slope, without trying to start again from particular points such as passes. From every pixel, an outflow on the model is simulated according to the nearest pixel of the line of larger slope.

3.6.3 Extraction of thalwegs by global outflow calculation or accumulation

Methods that follow exploit the same operative definition of thalwegs as techniques of progress previously presented: thalwegs, generated by phenomena of natural erosion of the ground, are places of river passage, assimilated to lines, understood this time as places where the pluvial waters concentrate. This definition is no longer in order to perform a local extraction of the network (either by local operators, or by following step by step a given line), but to exploit the whole surface to determine lines on which the strongest debits would appear in case of a uniform rain on the site.

The representation by a regular rectangular mesh is the one that leads to the simplest algorithms and will be the only one treated in this section (explanation data are even more strictly restricted to the square meshes). We will even speak of pixels (by reference to the terminology of Image Processing) to designate the points of the mesh.

In spite of these limitations, the methods present here the two interests of offering continuous line extraction and permitting some extremely simple implementations. As the results are very realistic on all land with a marked relief, these methods are among the most used.

3.6.3.1 Calculation of the drained surfaces

If we suppose there is a uniform rain across the whole of the land, the debit of the permanent regime is in any point proportional to the cumulative surface upstream of this point. Used in its discrete approximation, this property helps us calculate, for every pixel, its drained surface as the sum of the surface of a mesh (influence of the pixel considered) and the drained surfaces of pixels neighbouring, uphill, the considered pixel. Thalwegs are then local extrema places, on curves of constant altitude, of the drained surfaces. (See Figure 3.6.1.)

This definition gives rise to several algorithms, depending on whether one defines neighbourhoods in 4 or in 8-connexity, according to the way the relation 'being uphill of' is defined, a relation that we call relation of outflow, and according to the technique used to extract lines of thalwegs from the drained surface image.

We will restrict the explanation to the 4-connexity topology. The use of the 8-connexity, sometimes recommended, causes the problem of a incoherent topology: paths of outflows materialized locally by relations of neighbourhoods cross themselves, which doesn't correspond to a coherent physical model. Let us note that the techniques presented below are easily transposable in 8-connexity, maybe to the cost of an a posteriori correction procedure of resulting artefacts, which are always local.

The choice of the outflow relation leads to the distinction between two families of algorithms, presented briefly here. The techniques of extraction of thalwegs lines themselves, that can be chosen independently of the algorithm used for the calculation of surfaces, are presented in a different section.

The simplest approach consists in considering that every pixel cannot be uphill of more than one pixel among its neighbours. The downstream pixel will be chosen, on the one hand, so that its altitude is strictly lower

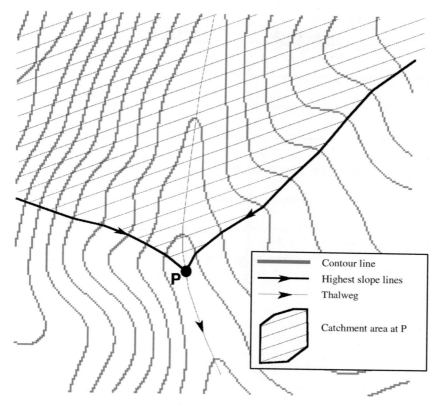

Figure 3.6.1 Catchment area of a point.

than that of the pixel considered, and on the other by a criterion of slope (the outflow of waters following the steepest slope).

This formulation ensures that the graph of the outflow relation is without closed cycle, in other words that this relation defines a tree on the DTM. (See Figure 3.6.2.)

The calculation of drained surfaces can be treated with a recursive algorithm or, more elegantly, using the natural order of altitudes to start the calculation with the highest points, and accumulating drained surfaces in one pass.

In the formulation of Le Men (Le Roux, 1992), the downstream pixel is the pixel for which the slope is steepest. This choice is operative on hilly landscapes (strong local variations of slopes), but produces drifts on smooth surfaces, bound to the quantification of outflow directions: the algorithm presents a defect of isotropy, a function of the direction of the mesh.

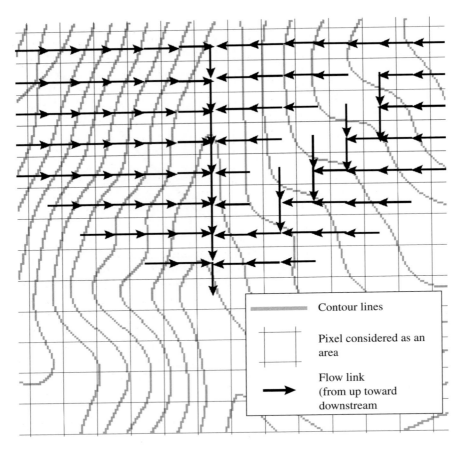

Figure 3.6.2 Tree of the outflow relation on a DTM.

The method of Fairfield (Fairfield and Leymarie, 1991) palliates this defect by proposing a random choice of the pixel downstream between the two pixels materializing the two larger slopes, the probability of choice of one of the two points being defined as proportional to the slope. This rule guarantees the identity of the mean direction of the pixel downstream and of the true direction of the steepest slope on a tilted plane, whatever is the orientation of the mesh. (See Figure 3.6.3.)

Calculation on the graph of outflow

The anisotropy (or the sensitivity to the direction of the mesh) resulting from the unique choice of the downstream pixel the steepest slope is the origin of a second family of approaches. Considering that the average direction of

Extraction of characteristic lines of the relief 263

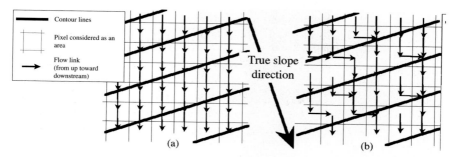

Figure 3.6.3 Outflow relation on a planar surface: (a) highest slope choice: the flow drifts away from the real direction; (b) two higher slopes random choice (Fairfield method): the average flow follows the real slope.

the outflow must remain faithful to the real direction of the steepest slope, Rieger (1992) proposed to take into account, as pixels downstream of a given pixel, all neighbours of lower altitude, and to propagate the surface drained of a pixel to its downstream pixels according to the slope values. In this formulation, the relation of outflow defines on the DTM a graph that is not other than the graph of neighbourhood oriented in the direction of decreasing altitudes (with exceptions of altitude equality, which we will return to). This oriented graph being without cycle, as the tree of outflow discussed previously, the algorithm of drained surface calculation can proceed in the same way (propagation of drained surfaces in the direction of decreasing altitudes).

This empiric formulation, which has the advantage of producing a more regular image of drained surfaces than Fairfield's method, does not, however, lead to an isotropic algorithm. A variant in 8-connexity due to Freeman (1991), founded on an ad hoc formulation of transmitted surfaces, leads to a better isotropy on surfaces of revolution, but not in the general case.

Jamet (1994) showed that, in the formulation of Rieger, the calculated drained surface is considered to be the calculated debit in the permanent regime, under a uniform rain on the whole DTM, and for a modelling in which pixels are assimilated to cells communicating by their lateral sides. For such a modelling, this debit depends on the orientation of the mesh in relation to the real slope of the ground. The simple calculation of the debit by unit of length orthogonal to the slope, deduced from the debit of the mesh and slopes to its neighbours, insures the exact isotropy of the result. Besides, on regular surfaces, this calculation of isotropic debit permits a rigorous definition of the convexity of the surface, the gradient of this debit being equal to the unit on the plane, lower on concave and greater on convex surfaces. (See Figure 3.6.4.)

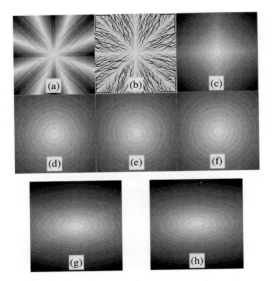

Figure 3.6.4 Drained surfaces computed on a spherical cap: (a) highest slope flow; (b) Fairfield method; (c) Rieger method; (d) Freeman method; (e) Jamet method; (f) theoretical expected result; (g) shows slight anisotropic effects on non-circular symmetric surfaces with Freeman method; (h) shows on the same surface as (g) the results of Jamet method.

Extraction of thalwegs

In all previous methods, the result of the phase of surface accumulation (or of calculation of the debit) leads to a map superposable on the DTM on which the researched thalwegs correspond to local extrema orthogonally to the calculated slope on the DTM.

Three types of techniques are used. The first relies on the classic techniques of image processing. Considering that the image of surfaces drained (or of the debit) is contrasted enough (or can become so after application of a digital skeletonization process), the binary mesh map of thalwegs is obtained by simple threshold of surfaces. The threshold corresponds to the minimal size of catchment area able to feed a drain. A vector representation of the network is deduced by a classic technique of vectorization (Le Roux, 1992). This method remains more suitable to approaches by outflow tree, as far as only these guarantee that the surface remains strictly increasing from the upper to the lower part of a drain (and therefore that the threshold won't be able to break their connectivity).

The second uses the tree of outflow as it has been defined previously, and remains therefore also more adapted to the corresponding family of methods. The tree of outflow is a vector structure linking the set of pixels of the DTM. The extraction of the thalweg network consists therefore

Extraction of characteristic lines of the relief 265

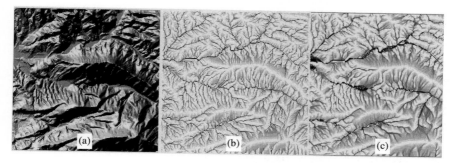

Figure 3.6.5 Drained surfaces computed on a DTM: (a) DTM (shaded); (b) highest slope algorithm; (c) Jamet's algorithm.

just in selecting a sub-tree of the outflow tree (Le Roux, 1992). The criterion of selection will be in the same way as previously a simple threshold of the drained surface (with the same guarantee of topological consistency).

The third constructs the vector representation of thalwegs in a recursive way, while starting with stronger debit drains. Except in an exceptional case, any thalweg goes either in another thalweg, or outside the treated zone. One can therefore count points of thalwegs' exit by carrying out an inventory of the local extrema of the surface drained on the edge of the DTM, then go up thalwegs following this local extremum uphill (until the drained surface goes below a threshold value). Recursively, new starting points can be detected along every thalweg, looking for local extrema of the gradient of the drained surface, which correspond to the junction of an incoming drain (Rieger, 1992).

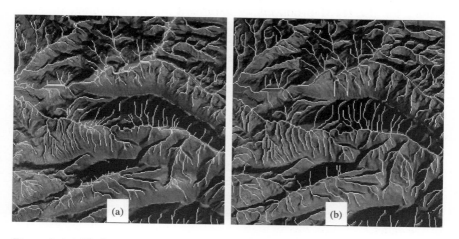

Figure 3.6.6 Thalweg extracted by the highest-slope method (a) and by Jamet's method (b) superimposed on a shaded DTM.

In any case, the obtained detections can contain artefacts. The too-short drains are therefore either ignored (Rieger, 1992), or eliminated a posteriori by too-short bow suppression in the obtained network or by morphological ebarbulage.

3.6.3.2 Considerations on the artefacts of the DTM

The above-described methods suppose on the one hand that outside of natural basins of the landscape and the sides of the site, any point of the DTM possesses at least a neighbour of lower altitude, and on the other hand that the direction of the ground slope is defined everywhere. This last condition being equivalent to the absence of a strictly horizontal zone, one can speak of an hypothesis of strict growth of the DTM between the natural outlets of thalwegs and the summits.

This hypothesis is never verified. The step of sampling data indeed limits the possibilities of representation of the present shapes on the surface, and in particular of the steep-sided valleys: so, along a thalweg crossing a valley whose width shrinks until a dimension commensurable with the sampling step, the altitude won't be able to vary in a monotonous way and one will observe on the DTM a local minimum of altitude upstream of the steep-sided setting. In the same way, the step of quantification in altitude fixes a lower limit to the representable slopes on the DTM: any weaker slope zone will appear as a strictly horizontal zone.

To these effects contingent to the discrete representation of data, some effects particular to the used source can be added: noise of acquisition on raw data generating local minima of the altitude, shortcomings of interpolation (for example, creation of horizontal zones for lack of information in the interpolation of contour lines), consideration in the representation of elements not belonging to the topographic surface (for example, vegetation, for the DTM produced by automatic image matching, that can cause an obstruction to a thalweg) ...

For grounding these artefacts, one supposes on the one hand that the ground doesn't include a strictly horizontal zone – their process will consist then in inferring the sense of the outflow from their environment – and on the other hand that only the important basins are meaningful – this notion is specified farther.

Faced with such problems, some authors have proposed solutions based on local techniques: these approaches result in giving an inertia to the outflow, which will permit it to clear zones of null slope, or even to go up the slope to come out of the artificial basins (see for example (Riazanoff et al., 1992) in the case of methods by progression). These methods don't permit an efficient control of the restored thalwegs' shape. We will limit the following, therefore, to methods based on a larger analysis of shapes of the relief.

Horizontal zone process

The horizontal zones can be solved in two manners: either one considers outflows as indeterminate on their extent, or one wants to define them.

In the first case, the horizontal zones are treated as lakes, fed upstream by the neighbouring pixels of larger altitude (that one will call input points), and opened on the neighbouring pixels of lower altitude (that one will call output points). Practically this option means calculating the relation of outflow, no longer on the topology of the usual neighbourhood induced by the mesh of the DTM (4 or 8-connex), but on a more elaborate topology, in which the nodes may be either points (pixels) or surfaces (the horizontal zones) – while relations of the neighbourhood remain relations induced by geometry.

At the end of the process of thalwegs extraction, the horizontal zones included in the tracing can be replaced by bows, whose geometry can be calculated, for example, by squeletization of these surfaces.

In the second case, the choice of the outflow sense on horizontal zones is subordinated to their shape and their environment. The simplest technique consists in forcing senses of outflow from input points toward output points. For that, one can consider for example that a pixel of a flat zone is uphill of one of its neighbours if and only if its distance to the nearest exit point is greater than that of this neighbour. The distance used can be a discrete distance, calculated by dilation of the exit border, constrained by the extension of the horizontal area. This approach can be completed by a consideration of the shape of the zone: considering that the thalweg (non-observable on the DTM) must cross the centre of the zone, one can for example subordinate directions of outflow to the whole of the border of the zone, that is to say simultaneously to the distance to input points and the distance to output points as well. This last option has the advantage of also being valid if the horizontal zone does not possess an output point.

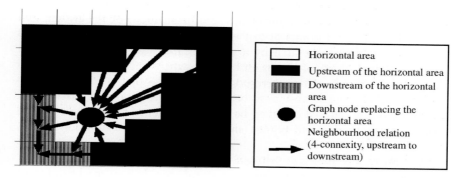

Figure 3.6.7 Flow topology induced by a flat area.

These types of calculation mean to accord a digital value to every pixel of the horizontal zone, which can be chosen in order to respect the required property of strict growth on the DTM. This set of values can then be assimilated to the decimal part of altitudes in calculations of slope.

Process of basins

One designates under the term of basin the catchment areas of the local minima of the DTM, which in most cases are correlated to artefacts. The correction of these artefacts consists in searching for one or many points of exit to the border of these catchment areas, and to force the outflow of the local minimum toward these points of exit (and therefore in the sense opposite to the slope).

The position adopted by most authors consists in minimizing the incoherence of the outflow, that is to say to search for the points of exit whose height over the local minimum is the weakest possible. All basins not being necessarily artefacts, this process will generally be controlled by a threshold on this height — or denivelate, the basins whose exit are higher than the threshold being preserved (that is to say considered significant).

The search for the points of exit of a basin cannot, however, be done independently of its environment: in complex situations, where numerous basins are neighbouring, the local determination of points of exit can lead to the simple fusion of neighbouring basins forming a new large-sized basin. If this process remains foreseeable while applying it recursively until disappearance of all non-meaningful basins, it can be very expensive. One has therefore rather to formulate the problem of the basin resolution like a search for a global minimum to the whole site. An economic solution — as it produces an algorithm of low complexity — consists in defining the total cost of the basin resolution as the sum of denivelates to cross in the sense inverse to the slope. Indeed, the sum of local minima altitudes being a constant of the problem, the minimization of the sum of denivelates is equivalent to the minimization of the sum of altitudes of the chosen exit points. The choice of points of exit is thus independent of the sense of the outflow, and can be operated by a simple search for the minimal weight tree in the dual graph of the basin borders, where each bow is valued by the minimal altitude of the corresponding border — that is to say the altitude of the existing lowest pass between each neighbouring basin pair. The sense of the outflow can then be calculated recursively in the obtained tree, from the known exit points (sides of the site or meaningful basins). (See Figure 3.6.8.)

Once the output points are chosen, as in the case of the horizontal zones, the process of basins can give rise to a simple topological modification or to a geometric process of the DTM.

Extraction of characteristic lines of the relief 269

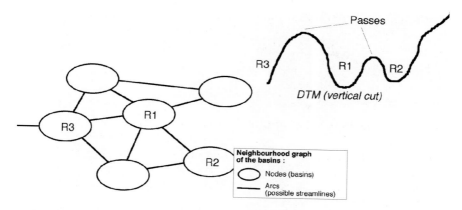

Figure 3.6.8 Search of the basin exit: the simple choice of the lowest pass leads to the fusion of R_1 and R_2 without giving them an outlet. A global search has to be performed in the graph of neighbourhood of the basin.

Among the topological processes, the two most obvious techniques consist in adding a relation of outflow between the local minimum of every basin and its pass of exit – with the drawback of producing non-connex thalwegs at the end of the process – either to reverse the sense of the outflow on pixels leading from the local minimum to the pass of exit following the steepest slope (Fairfield and Leymarie, 1991). This last technique can however be used only in the case of accumulation calculations on the outflow tree: the inversion of the sense on only one path creates indeed some cycles on the graph of outflow, and does not permit the other techniques to be used any longer.

The most current geometric process consists in modifying the geometry of the DTM to construct an exit path of constant altitude, by searching for a pixel of lower altitude than the local minimum downstream of the exit pass, by levelling of the DTM on the path of steepest slope joining these two pixels (Rieger, 1992). This last solution is more adapted in the case of accumulation techniques on the entirety of the graph of outflow.

Let us note finally that the order of process of the horizontal zones and the basins is not irrelevant: as the process of basins is leading to a modification of outflows on the DTM, it must evidently be applied before the process of horizontal zones.

3.6.3.3 *Annex results*

The methods presented in this section have the advantage of giving access to a description of a topography richer than the simple network of

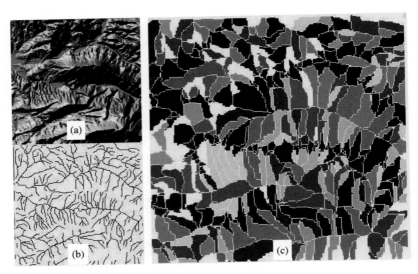

Figure 3.6.9 Watershed extracted by an outflow method: (a) DTM (shaded); (b) thalweg network; (c) watersheds (represented by various grey levels, and their boundaries in white).

thalwegs. The calculation of the horizontal zones and basins, understood as artefacts of the DTM, provides an indicator of its quality. The calculation of the drained surface offers a natural hierarchy of the network of thalwegs, which can prove to be useful in numerous applications. The materialization of outflows on the whole surface permits the direct calculation of catchment areas associated to each thalweg or every portion of thalweg (for example, by simple labelling of its pixels uphill in the outflow tree). This decomposition of the surface is used extensively, in particular for the physical modelling of flood phenomena (Moore *et al.*, 1991), but also for the cartographic generalization (Monier *et al.*, 1996). Finally, the network of crests, subset of catchment areas borders, is also accessible by this type of technique, either that one adapts algorithms of accumulation to calculate a function of adherence to catchment areas on the same mode as the calculation of the drained surfaces, or that one uses directly calculated catchment areas to extract their borders by techniques of vectorization.

3.6.4 Extraction of thalwegs from local properties

This family of methods was historically the first to be used (Haralick, 1983). For this algorithm and those that followed, the common idea is to search, on every pixel of the DTM, independently of results on the neighbouring pixels, to see if the current point verifies a certain relation with its environment; in which case, it receives a stamp that identifies it as a point of crest or thalweg. Only some methods, considered as representative, will be presented below.

3.6.4.1 By analysis of the discrete representation

The surface is analysed here in a discrete way. Kim's method (1986) is the more often evoked. The algorithm proceeds by a scan of the DTM according to X and Y in order to identify elementary geomorphologic shapes.

With this method, the algorithm scans the DTM from left to right, then from top to bottom. During the sweep, characters are accorded to pixels (of the DTM) successively according to the nature of the slope that is between them. A threshold of altitude differences is applied to define the horizontal surfaces. Transitions between points are qualified according the following rules:

- the passage of a point to a higher one is represented, e.g. by the character '+';
- the passage of a point to a lower one is represented by the character '−';
- the passage without change of the level of a point to a higher one is represented by the character '='.

The interpretation of the shapes of models (images constituted by the three characters +, − and =) produced by two scans makes possible a first analysis, according to the scan profile, of the local morphology of the ground. The analysis relies on successions of slope characters '+', '−', '=', for example, if a '−', is followed by a '+', it indicates the presence of a steep-sided thalweg, while several '−' followed by several '=' followed again by one or several '+' could mean a large valley bottom or the flat bottom of a basin.

The superposition of the two scan models, in the longitudinal and transverse sense, permits one by comparing the information, to deduce characteristics of the local morphology in the two dimensions.

Thus the obtained model allows, after an ad hoc filtering, elements of the main characteristic lines network to be recovered.

In certain cases, shapes are not very clean, in particular in the flat zones. In addition, the connection of the network is not always satisfactory, a certain number of segments remaining isolated.

3.6.4.2 By analysis of a continuous representation

The surface is assumed to be continuous, or continuous by pieces so as to provide an analytic expression of the surface. The function representative of the altitude must make it possible to give an account of the different discontinuities due to the terrain:

- slope, cliffs directly bound to the altitude;
- lines of slope rupture (concavity, convexity);
- line of curvature rupture.

On neighbourhoods of given size, the continuity of the slope can be obtained by junction of parabolic bows and that of the curvature by the utilization of cubic bow. Several methods explored this possibility; Haralick (1983) makes the analysis with the help of a model using some bi-cubic functions. The method that will be presented (Dufour, 1988) relies on the hypothesis that the ground can be modelled with the help of a Taylor series, which will be limited here to the order 2, which permits one to describe a sufficiently large number of phenomena.

$$Z = H(X,Y) = H(X_0, Y_0) + ax + by + \tfrac{1}{2}(cx^2 + 2dxy + ey^2) + \varepsilon,$$

where $x = X - X_0$ and $y = Y - Y_0$ are considered small.

Formulae of determination of coefficients of the polynomial depend on the chosen mesh. The square configuration will be used here and the coefficients, calculated by the least squares method are expressed in the following manner (Dupéret, 1989):

$$H_0 = \frac{5H_0}{9} + \frac{2}{9}(H_2 + H_4 + H_6 + H_8) - \frac{1}{9}(H_1 + H_3 + H_5 + H_7)$$

$$a = \frac{1}{6}(H_1 + H_7 + H_8 - H_3 - H_4 - H_5)$$

$$b = \frac{1}{6}(H_1 + H_2 + H_3 - H_5 - H_6 - H_7)$$

$$c = \frac{1}{3}(H_8 + H_4) - \frac{2}{3}(H_2 + H_6) + \frac{1}{3}\left(H_1 + H_3 + H_5 + H_7 - \frac{2}{3}H_0\right)$$

$$d = \frac{1}{4}(H_1 + H_5 - H_3 - H_7)$$

$$e = \frac{1}{3}(H_2 + H_6) - \frac{2}{3}(H_8 + H_4) + \frac{1}{3}\left(H_1 + H_3 + H_5 + H_7 - \frac{2}{3}H_0\right),$$

where H_i is defined according to Figure 3.6.10.

The horizontal curvature is a very interesting variable, which uses all first and second derivatives of the altitude function. It takes very strong negative values in thalwegs, remains weak in the regular sides and becomes very strongly positive along the crests.

The horizontal curvature at point $(x, y) = (0, 0)$ of contour lines is expressed according to the coefficients above:

Extraction of characteristic lines of the relief 273

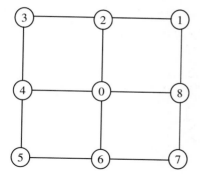

Figure 3.6.10 Numerotation of the points of the neighbourhood, of altitude $H_0, H_1 \ldots H_8$.

$$\Gamma_N = \frac{2abd - cb^2 - ea^2}{(a^2 + b^2)^{3/2}}.$$

The same principle is applicable to the calculus of all variable geomorphometric types expressed according to derivatives of the altitude function: slope, curvature of level lines, curvature of slope lines, Laplacian, quadratic mean curvature, total curvature, etc.

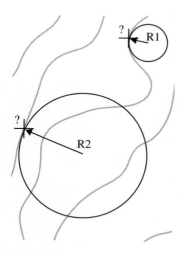

Figure 3.6.11 Curvature radius $R = 1/\Gamma_N$ in two points of the ground.

Figure 3.6.12
Representation of Γ_N (from the darkest to the lightest grey, the thalwegs, the crests, then the intermediate zones) with the superimposition of contour lines.

 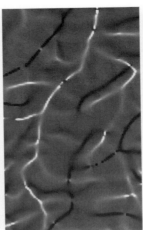 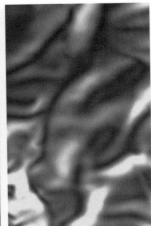

| Shading of the surface with a lighting to the NW | Representation of the horizontal curvature of contour lines | Representation of the slope for the same surface |

Figure 3.6.13

3.6.4.3 Synthesis

The search for points of the characteristic lines independently from each other eludes the very notion of characteristic line. If the acquisition mode and the density are appropriated, then these methods can provide a posteriori these lines. However, in the main, the extracted network is non-connex with sometimes segments of non-negligible thickness. Even if they are very robust, these algorithms very rarely restore completely the logic of a hydrographic network. The identified points often form dispersed segments, therefore non-connex, and these sets of points have a variable thickness.

Some mixed methods were also set up. Thus, the algorithm of Ichoku (1993) proceeds by combination of Kim's and Riazanoff's algorithms and provides a hierarchized and reasonably connected network. Basins are eliminated according to the principle of the inversion of the sense of the drainage. Although the general connection of the network is superior to the two basic methods used, some parasitic networks subsist.

3.6.5 Numerous applications

Examples that follow are by no means exhaustive. They are mentioned merely to show that some varied applications can be built on data produced by the above mentioned methods. Some do not require the connection of segments that compose the network. The mastery of the representation of characteristic lines network is the basis of various measures on watersheds.

3.6.5.1 The hierarchization of a hydrographic network

The networks of thalwegs produced by different methods can allow the hydrologists to build a hierarchical classification of segments composing the detected network. Two methods among all those possible are presented here.

The first use by Riazanoff is introduced by Shreve (1967). A segment is a part of the active network of a point source to a confluence, or of a confluence to another confluence. Every segment possesses a hierarchical value that is the sum of the hierarchical values of immediately upstream segments, segments that are the most upstream (descended of a source) receiving the hierarchical value 1. (See Figure 3.6.14.)

A second method, due to Horton (1945) and improved subsequently, establishes the hierarchy in the following way:

- rivers not having even received an affluent are of order 1;
- rivers having received at least a river of order 1, therefore 2 rivers of order 1 are of order 2;

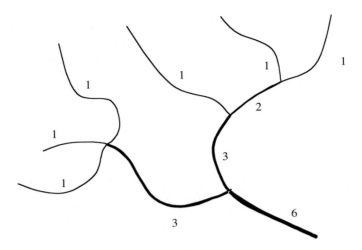

Figure 3.6.14 Example of network in the classification of Shreve.

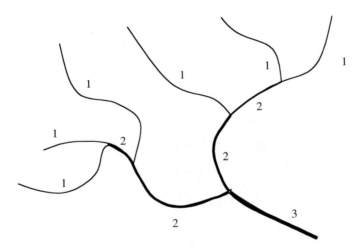

Figure 3.6.15 A similar network in the classification of Horton.

- rivers having received at least a river of order 2, therefore 2 rivers of order 2 are of order 3;
- and so on.

(See Figure 3.6.15.) The final order of the outlet river of a catchment area has a comparable value from a basin to the other, provided that they result from surveys of the same scale.

It is then possible for the user to establish relations of comparison between the order, the designation of the river and the surface of catchment areas. For example, the denomination 'large river' will be reserved to segments of order 7 and 8, with associated catchment areas of a surface between 10,000 and 100,000 km^2.

With the help of this hierarchization of networks, the user can also use thresholds on the order of segments in order to limit some parts of the drainage network.

3.6.5.2 *The cartographic smoothing*

Despite all precautions, the filtered DTM presents softer ground shapes in comparison to what they should be. To limit the deterioration of terrain shapes, one may therefore use a cartographic smoothing process that has to determine as automatically as possible the zones of the DTM that correspond to the characteristic lines of the ground.

For example, the program used in IGN-F calculates a digital model of horizontal curvature in the DTM (performing a classification in five domains) with homogeneous geomorphometric properties describing highly convex zones (crests) up to strongly concave zones (thalwegs).

Parameters of adapted smoothing can then be applied in an uniform way on each of its zones to get the contour lines smoother, but preserving a good description of the terrain shapes. (See Figure 3.6.16.)

Figure 3.6.16 Comparison of level curves restored by hand (in grey) and cartographically smoothed (in dark grey).

3.6.5.3 Automatic cut-off of watersheds

As an indispensable component to the rational management of digital data in GIS, the limits of watersheds has for a long time been digitized by hand before being inserted digitally. The first studies in automation started in 1975. One of the first methods (Collins, 1975) proposes to classify all points of the DTM by increasing order of altitude. The lowest point is an outlet of the watersheds; if the second lowest point is not connex to it, it means therefore that it belongs to another basin. The provided results are good but are unstable with a real DTM where objects on the ground, the plane zones and irregularities of thalwegs' profiles destabilize the procedure of detection.

Since then numerous methods have been studied, trying to extract from the DTM the points presenting the property of adherence to a line of crest. The different strategies used to make the set of points connex and of unit thickness have most of the time stumbled on the discontinuous characters of lines obtained and the problem of local minima. The assimilation of the main and secondary crest lines in methods of non-hierarchical detection puts the problem of identification of the real lines of water sharing. The theory of the mathematical morphology constituted an efficient and original contribution to this type of survey while permitting an automatic method of catchment areas cut-off to be finalized (Soille, 1988).

3.6.5.4 Manual editing of DTM

The characteristic line extraction as it has just been presented is often disrupted in practice by the presence of micro-reliefs, artefacts of calculation, as small crests, passes or aberrant basins, without physical reality.

The elimination of these topographic anomalies can be performed by replacing the series of elevations, in the considered thalweg, by a new set of data, obtained by a controlled diminution of the values. It is easy but poses the problem of other altitude modifications to the neighbourhood of the thalweg. The step of interactive manual correction proves to be, for the meantime, still indispensable. On its duration depends the economic opportunity to make these corrections by hand. At best, three DTM editors are accessible in some software:

- A punctual editor, intended to modify only one altitude at a time (used rarely).
- A linear editor working in such a way that the operator indicates interactively a 2D or 3D axis with a chosen influence distance. Inside the concerned zone, altitudes are resampled according to hypotheses made on the desired profile of ground perpendicularly to the axis indicated by the operator (profile in U, in V, linear, bulldozer . . .). This editor is useful for suppressing altimetric prints of hedges in the DTM, or when characteristic lines and/or a hydrographic network is available.

Extraction of characteristic lines of the relief 279

Figure 3.6.17 In black: example of linear object that can give rise to updating of the profile of the ground in the DTM among the four possibilities and distance of interpolation (*D*) to the axis of the object: uniform interpolation, in *U*, in *V* or some in bulldozer. Surfacic processes can be applied to objects systematically known by their perimeter: a clear hydrographic surface (in the bottom-left corner, surrounded by a clear line) can be given a rigorously constant altitude, a forest (right, surrounded by a clear line) can receive a bias corresponding to the mean height of the trees.

- A surfacic editor that allows the operator to select a process to apply to altitudes situated inside a polygon: various resamplings, force to a constant level, application of a bias, parametrable filtering of the objects over the ground . . . .

The manual correction operations are shorter and more efficient if the DTM quality is good. At this step, in spite of everything, the user will not avoid some topographic anomalies, like the thalweg perched in the convex zone, the crest enclosed in a concave zone or the delta of a river in which the stream passes from convergent to divergent zones.

3.6.5.5 Toward a partition of the ground in homogeneous zones?

The different adopted definitions lead to a partition of the domain in catchment basins that must be completed to achieve morphologically a separation in homogeneous domains.

The decomposition of the ground in homogeneous zones under shape of curvilinear triangles is foreseeable. It represents immediately a finality for the topographer and a way toward a rational generalization of the relief. The junction between two neighbouring triangles takes place along a sinuous curve corresponding to singularities of the function altitude in general. Such a model is attractive because it inserts without difficulties delicate topographic situations such as:

- valleys with a low slope with a sinuous course, and
- rocky-toothed cliffs,

which required a decomposition of the domain into numerous facets.

Every curvilinear triangle includes a parametric representation with a limited number of parameters (e.g. 30 coefficients for a parametric system of cubic type). The value taken by these functions in the border of zones makes it possible to recover the curvilinear character of separation lines.

In the curvilinear domain (ABC), X, Y and H are represented by cubic functions of the barycentric coordinates of the ABC triangle. Cubic functions $X(\alpha, \beta, \gamma)$, $Y(\alpha, \beta, \gamma)$, $H(\alpha, \beta, \gamma)$ include each ten independent parameters that can be defined by values of (X, Y, H) to the three summits, as well as in two points on every curvilinear side and one central point.

$$\alpha = \frac{\text{area (MBC)}}{\text{area (ABC)}} \quad \beta = \frac{\text{area (MAC)}}{\text{area (ABC)}} \quad \gamma = \frac{\text{area (MBA)}}{\text{area (ABC)}}$$

and

$$\alpha + \beta + \gamma = 1.$$

Separation lines are to be looked for giving priority to where there are obvious discontinuities:

- of the function altitude (slope, cliff);
- of the derivative of the slope (crest, steep-sided thalweg);
- of the curvature (slope rupture).

The convenient exploitation of such a model presents various difficulties whose major problem is the reliable automatic determination of lines of homogeneous zone separation.

3.6.5.6 Complementary approach

The automated extraction of the characteristic line networks is of course a privileged way of knowing the surface of the land, which comes in complement to other approaches, for example: the determination of surface envelopes (Carvalho, 1995). A surface envelope is a general, schematic and theoretical model of the relief that makes it possible to recover, in certain cases, the general lines of a situation of maturity of the relief, and to analyse the links between the present and old topographies. The surface envelope relies on the high points of the relief of a region, so as to eliminate irregularities of the topographic surface coming from the linear erosion of rivers. Indicators of surface thickness either eroded or erodable can be deduced from differences between DTM and the surface envelope to provide some useful elements to the evaluation of erosion balances. The automatic detection of summit points is considered here therefore like a tool serving the development of a numeric representation of surface envelopes while giving points of passage. It is followed by a phase of construction of the surface. The practice shows that the automation of this determination is justified when the DTM mesh is adapted to the land. The simultaneous knowledge of networks of thalwegs and surface envelopes allows the imperfections of the thalwegs network in relation to the summit surface envelopes to be determined (direction of the thalwegs network not compliant with the lines of steepest slope of the surface envelopes).

References

Carvalho J. (1995) Extraction automatique d'informations géomorphométriques (réseaux et surfaces enveloppes) à partir de modèles numériques de terrain. Ph.D. Thesis, Université de Paris 7.

Chorowicz J., Parrot J.F., Taud H., Hakdaoui M., Rudant J.P., Rouis T. (1995) Automated pattern-recognition of geomorphic features from DEMs and satellites images Z, *Geomorph. N.F.* (Berlin), Suppl. Bd101, pp. 69–84.

Collins S. (1975) Terrain parameters directly from a digital terrain model, *Canadian Surveyor*, vol. 29, no. 5, pp. 507–518.

Depraetère C., Moniod F. (1991) Contribution des modèles numériques de terrain à la simulation des écoulements dans un réseau hydrographique, *Hydrol. Continent.* vol. 6, no. 1, pp. 29–53.

Dufour H.M. (1977) Représentation d'une fonction par une somme de fonctions translatées, *Bulletin d'information de l'IGN*, no. 33, pp. 10–36.

Dufour H.M. (1983) Eléments remarquables du relief – définitions numériques utilisables, *Bulletin du comité français de cartographie*, no. 95, pp. 57–90.

Dufour H.M. (1988) Quelques idées générales concernant l'établissement et l'amélioration des Modèles Numériques de Terrain, *Bulletin d'information de l'IGN*, no. 58, pp. 3–18.

Dupéret A. (1989) Contribution des MNT à la géomorphométrie, *Rapport de stage*, DEA SIG, IGN – IMAGEO.

Fairfield J., Leymarie P. (1991) Drainage network from grid digital elevation models, *Water Resources Research*, vol. 27, no. 5, May, pp. 709–717.

Freeman T.G. (1991) Calculating catchment area with divergent flow based on regular grid, *Computer and Geosciences*, vol. 17, no. 3, pp. 413–422.

Haralick R.M. (1983) Ridges and valleys on digital images, *Computer Vision, Graphics and Image Processing* 22, pp. 28–38.

Horton R.F. (1945) Erosional development of streams and their drainage basins: hydrophysical approach to quantitative morphology, *Geological Society of America Bulletin*, vol. 56, pp. 275–370.

Ichoku C. (1993) Méthodes automatiques pour l'analyse et la reconnaissance des systèmes d'écoulement en surface et dans le sous sol. Thesis, Université de Paris 6.

Jamet O. (1994) Extraction du réseau de thalwegs sur un MNT, *Bulletin d'information de l'IGN*, no. 64, pp. 11–18.

Kim Y.J. (1986) Reconnaissance de formes géomorphologiques et géologiques à partir de modèles numériques de terrain pour l'exploitation de données stéréoscopiques Spot. Ph.D. Thesis, Université de Paris 6.

Le Roux D. (1992) Contrôle de la cohérence d'un réseau hydrographique avec un modèle numérique de terrain, *Rapport de stage*, Laboratoire COGIT, IGN, Saint-Mandé, France.

Le Roux D. (1993) Modélisation des écoulements sur un modèle numérique de terrain – Applications aux crues et inondations du 22/09/89 à Vaison la Romaine, *Engineer End of Study Memoir*, ESGT, Le Mans, France.

Monier P., Beauvillain E., Jamet O. (1996) Extraction d'éléments caractéristiques pour une généralisation automatique du relief, *Revue Internationale de Géomatique*, vol. 6, no. 2–3, pp. 191–201.

Moore I.D., Grayson R.B., Ladson A.R. (1991) Digital terrain modelling: a review of hydrological, geomorphological and biological applications, *Hydrological processes*, vol. 5, pp. 3–30.

Riazanoff S. (1989) Extraction et analyse automatiques de réseaux à partir de modèles numériques de terrain. Contribution à l'analyse d'images de télédétection. Ph.D. Thesis, Université de Paris 7.

Riazanoff S., Julien P., Cervelle B., Chorowicz J. (1992) Extraction et analyse automatiques d'un réseau hiérarchisé de thalwegs. Application à un modèle numérique de terrain dérivé d'un couple stéréoscopique SPOT, *International Journal of Remote Sensing*, vol. 13, pp. 367–364.

Rieger W. (1992) Automated river line and catchment area extraction from DEM data, *ISPRS Congress Commission IV*, Washington DC, August, pp. 642–648.

Shreve R.L. (1967) Infinite topologically random channel networks, *Journal of geology* (Chicago), vol. 75, pp. 178–186.

Soille P. (1988) Modèles numériques de terrain et morphologie mathématique: délimitation automatique de bassins versants, *Engineer End of Study Memoir in agronomy, orientation 'génie rural'*, Université catholique de Louvain la Neuve, Belgium.

3.7 FROM THE AERIAL IMAGE TO ORTHOPHOTOGRAPHY: DIFFERENT LEVELS OF RECTIFICATION

Michel Kasser, Laurent Polidori

3.7.1 Introduction

The very principle of an image acquisition, which means a conical perspective for specialists, implies that a photographic image is not generally superimposable to a map. It is partly due to the existing reliefs in the observed object, and to the fact that even for an object that would be rigorously plane, the optical axis has no reason to be precisely perpendicular to this plane. For a very long time, techniques producing the controlled distortions of photographs have been developed in order to give these images a comparable metrics to that of a map. These techniques can be conceived at very elaborate levels, according to the requested quality. Rigorously speaking, only when the correction is perfect is the product obtained called orthophotography, but there are many intermediate solutions, that one uses to get rectified images. Here we aim to clarify the different levels possible, which have some very different costs, and corresponding large differences of precision.

3.7.2 What is an orthophotography?

The orthophotography is a picture (generally aerial) that has been geometrically rectified to make it superimposable in any place on a map, possibly with enriched graphic additions. These additions can originate either from external information (administrative limits, toponyms, etc.), or from the interpretation of the image itself (in order to ease its use: drawing of roads, of building contours, etc.). If one compares it with classic cartography, the orthophotography differs by the absence (that can be total) of most phases of interpretation and drawing. As such tasks always require an important human added value, the possibility of suppressing them permits therefore an almost total automation of the process: it is through the considerable reduction of costs and delays obtained that digital orthophotography became a product more and more current, capable of completely replacing the traditional cartography.

The process of manufacture necessarily implies knowing two data sets: parameters of image acquisition (orientation of the image, spatial position of the camera) and the relief (described generally by a DTM, digital terrain model). It is therefore clear that the precision of the final product depends directly on the quality of these two data sets.

Let us note here that one knows how to get a DTM currently by automatic image matching of images. It is more or less today the only operation that may be completely automated in digital photogrammetry, as the other tools that tomorrow may be accessible (automatic drawing of roads, extraction of buildings, etc.) currently require the supervision of a human operator.

The user will sometimes not need a very advanced cartographic quality, only a fairly constant mean scale, for example within 10 per cent, which is not compatible with the raw photo but may be performed without rigorous photogrammetric process. For example, commercial software for image manipulation (advertisement, editing, retouching of amateur photographs, etc.) may be used the proper way: one will search for some identifiable points of known coordinates, and one will distort the image in order to oblige it to respect their positions. Certainly it is not at all an orthophotography, but it may provide some facilities and will often be less expensive. But it is necessary to avoid using the same denomination: one will speak, for example, of rectified images. The term 'orthorectified' image or orthophotography would be an abuse here. We will use the generic term 'rectified' image for any image that underwent a process that distorts it geometrically to bring it closer to an orthophotography. This rectification can therefore be performed at various levels: so the orthophotography represents the highest possible level concerning rectification.

Otherwise, most users need numerous images to cover a given area, which requires a 'mosaiquage', meaning that it is necessary to process the links between images so that one will no longer notice them, in terms of geometry (breakage in linear elements) and of radiometry as well (differences of grey level, or of hue, for one given object on two nearby images). For the geometric aspects, the previous considerations will help to get an idea of the nature of the problems susceptible of being met. On the other hand, for radiometric aspects, the physical phenomena originating them are numerous: chemical process of images, difference of quantity of light distributed by the same object in different directions (bidirectional reflectance), images possibly acquired at different dates, or merely at different instants (and therefore lighting), etc.

Let us add again a few elements of terminology used in the spatial images process:

- One can say that a image is 'georeferenced' if a geometric model is defined, permitting one to calculate the ground coordinates of every point of the image. For example, a grid of distortions will be provided with the image, this remaining raw otherwise.
- An image is said to be 'geocoded' if it has been already more or less rectified (with, in consequence, a better scale constancy).

3.7.3 The market for orthophotography today

One will have understood that all these photographic products have in common to leave the work of interpretation to the final user. Experience shows besides that, if most users of geographical information have in general sometimes important difficulties in reading and exploiting a map, on the other hand practically everybody will be able, even without any technical culture, to understand and to interpret the full content of a photographic document. Multiple cadastral investigation examples in rural Africa can be found in support of this observation.

This ease of use makes it in fact an excellent support of communication, or even of negotiation, in particular for decision making related to regional development (individual citizens, elected people, administrations, etc.).

Otherwise, as mentioned previously, the development of rectified images is capable of a very high level of automation, which has considerably decreased the production costs for some years. One even foresees shortly an entirely automatic production possibility, with very common computers of PC-type. One will thus have access to a product hardly more expensive than the image of origin (this is more or less true as far as the DTM is available), and that will be able to be practically orthorectified without additional cost. Besides, it could become of little use to produce badly rectified images.

Let us note again that the recent evolutions of the PC permit comfortable manipulations of digital images, even the very voluminous and numerous ones, so that the orthophotography often became the basic layer of urban GIS, the only one that may be updated regularly.

There are therefore many reasons to anticipate a major development of the market of the digital orthophotography, and it will not necessarily be the same for imperfectly rectified images.

Besides, several countries have undertaken or even finished and maintain, a national coverage in digital orthophotographies.

3.7.4 The different levels of rectification

Given that heterogeneous metric quality products discussed previously coexist, it appears desirable to clarify concepts of the used terminology.

There are only five applicable quality criteria on the subject, if we look at the customer's needs:

1. *Geodetic criterion*: precision of the correspondence between coordinates of objects in the official national reference frame and in the rectified image (if coordinates are displayed in the document): so, if the system of coordinates is not sufficiently known (for lack of links to the local geodesy for example), the coordinates that the user can extract from the image will be in error by a constant value, which can be very significant. This type of error, for certain applications, has no importance, but will be bothersome in others (e.g. for somebody performing further work with GPS).

2. *Scale quality criterion*: stability of the image scale, or more precisely evaluation of the scale gap between the image and the corresponding theoretical map (whose scale is not precisely constant according to the system of projection used). One can speak of a scale stability within 10 per cent (coarse process), and up to 0.01 per cent (the most precise processes). This criterion is certainly the most important of all. Let's not forget here that the precision of the DTM (a concept otherwise difficult to define exactly) appears directly in this criterion: if the DTM is false, the rectification will necessarily be the same!

3. *Object scale criterion*: in relation with the criterion (5) below, does the criterion (2) apply only to the ground, or also to objects above the ground? This point is currently important because the DTM only concerns the ground, and removes, as much as possible, these objects (buildings, trees, etc., see §3.3).

4. *Radiometric process criterion*: concerning the visual quality of the rectified image, we must note that all resampling will imply a certain

deterioration of the contour cleanness. The visual quality of the image assembly includes several more or less appreciable aspects: specular reflections on water surfaces (whose suppression sometimes implies heavy manual work), balancing of hues on links (that may require a human intervention for a good result), etc.

5. *Building process criterion*: in the case of the urban zones, the presence of large buildings creates specific difficulties often implying a manual intervention (digital model of the relief is incomplete because of zones that are not seen in stereo, hidden zones where images' sources do not give any information, etc.). In this criterion, two important technical aspects are the value of the focal length of the camera used (the more important it is, the less the image acquired is usable to get the DTM, so that it will be necessary with the use of a long focal distance to have the DTM already done elsewhere, but the more the problems of building facades will be reduced), and the overlap between the exploited images (if the overlap is very significant, it allows one to use only the central zones of images, where there are little or even no problems with buildings).

For these last two aspects, the operator's interventions can be more important. The final quality is thus quite difficult to specify, and it implies some very variable and sometimes major costs.

Finally one will be able to refer also to process specifications in use in the spatial imagery. Levels of SPOT images processes are typical examples of what is currently proposed:

- Level 1: raw image.
- Level 2: rectified image with the exclusive use of image acquisition parameters, but without information on the relief (product being sufficient in zones of plane for some applications). Such a product will be accessible when one obtains the digital images with their orientation and localization parameters.
- Level 3: image rectified by using image acquisition parameters and a DTM. The result will be therefore an orthophotography.

But currently, no terminology has been dedicated by the use to distinguish the different qualities of process. It would be most helpful if the rectified products suppliers systematically provide evaluations on the five criteria above. For the development of this sector of activities, it is indeed not desirable that the present confusion persists. For example criteria (2) and (3), although quite essential, are almost never specified today.

3.7.5 Technical factors seen by the operator

Let us now make a synthesis of elements of the technical process concerning the operator in the final quality of a rectified image:

- The DTM can be obtained directly from the exploited stereoscopic cover and the corresponding stereopreparation. *Criteria: precision of the stereopreparation, quality of the altimetric restitution.* If the DTM is provided by the customer, or obtained in available external databases, one will observe a separation of responsibilities between the operator and the supplier of the DTM, which can lead to a complete removal of responsibility from the operator! *Criterion: precision of altitudes of the DTM.*
- Parameters of image acquisition are, in all present processes, unavoidable by-products of the photogrammetric part of the process (aerotriangulation). *Criterion: are the parameters of image acquisition used?*
- The knowledge of the reference frame is not a simple problem. It even occurs in certain regions of the world that one doesn't have access to it at all. Without reaching this point, there is again there a sharing of responsibilities between the operator and the supplier of reference points, capable of removing the responsibility from the operator. Otherwise, the altimetry being generally referenced on the geoid and planimetric coordinates being purely geometric (referenced on an ellipsoid), a large confusion often reigns in altimetric reference systems used (national levelling network, GPS measures, etc.). Besides, traditional national planimetric networks have, at the time of a generalized use of the GPS, important errors often hardly known: one cannot reference an image rectified in such a system without adjusting the error model locally, and it implies ground measurements. *Criterion: knowledge of the planimetric and altimetric reference frames.*
- The human intervention is more critical to reach the desired quality (in particular visual): study of reference points, geometric continuity across the links, radiometric continuity, process of buildings, etc. *Criteria: qualifications, practice and care of the operator.*

It is finally necessary to put users on guard against the fact that more and more commercial software proposes the functionality of geometric distortion of an image that can look like a rectification of the image. It is due to the explosion of the market of the digital photograph (warping, morphing, etc.). As we saw previously, these processes that are completely unaware of the geometric origin of the treated problem can only provide some mediocre results. At a time where a rigorous orthorectification is becoming almost entirely automatic and therefore financially painless, we can only encourage the users to take care with such products.

3.7.6 Conclusion

The customer sometimes accepts too easily the available product limits for lack of a sufficient culture in geographical information. A clarification of the nomenclature of products and the underlying concepts is absolutely necessary if one wants to enjoy the expansion of this market, which must improve on healthier bases. Otherwise a certain 'democratization' of orthorectified images appears indisputable today. If beneficiaries of service in this matter don't come from the geomatics technical environment, and don't have a sufficient culture concerning geographical information (in particular concerning photogrammetry), it may be anticipated that the available products will for a long time present a quality level much lower than what is possible, this despite the perfectly good faith of the operator, and for the same price as a rigorous process ...

3.8 PRODUCTION OF DIGITAL ORTHOPHOTOGRAPHIES
Didier Boldo

3.8.1 Introduction

In theory, a line of orthoimages production can be analysed in very simple modules. It uses in input the images, the geometry of the image acquisition and a digital surface model (DSM). From these data one may build the orthoimages covering the whole zone. One calculates a mosaic, does a radiometric balancing and then archives the results. (See Figure 3.8.1.)

Images can originate from numerous sources. Generally, one uses very precisely silver halide scanned images (pixel of 21 to 28 microns). Their main inconvenience is their lack of radiometric consistency that creates problems at the time of the orthoimages mosaics calculation. These images generally have some very important sizes (typically of the order of 300 Mo for colour images), which requires computer power for their manipulation. If these data are used, the line of production must foresee some professional scanners, as well as the companion equipment.

But new types of data become available: images originating from digital cameras, from linear CCD aerial sensors and high-resolution satellite images. These data have the advantage of an excellent internal radiometric

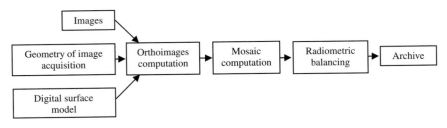

Figure 3.8.1 Production of digital orthophotographies.

consistency, which greatly simplifies the processes of mosaic calculation. However, these processes can represent an important part of the calculation of the orthoimage, notably in terms of operator time.

These images possess a geometry, produced through aero- or spatiotriangulation. These calculations allow one to determine the direction of every pixel of each image. These calculations can be integrated in the line productions. This situation has the advantage of allowing one to use algorithms in order to automate part of the work, e.g. the detection of link points (see §2.4).

The last requested data is the digital surface model. In the case of classic orthoimages, it is a digital terrain model that is generally available on the shelf as outside data in most western countries. But if one wishes to make a true orthoimage, it is necessary to use a more precise model. This model can come from an aerial laser scanner, from automatic correlation, or from classic photogrammetric restitution. If these models are precise and corrected enough, they can even permit one to straighten the images of the buildings.

In fact, the necessary data to manufacture orthoimages can be integrated in assembly lines of production. So the aspects of calculation of the image geometry (see §1.1), of calculation of the DSM (see §3.2), and problems of colour consistency (see §1.3) are integrated generally in the production chain.

Some automatic tools for the calculation of join-up lines and for balancing radiometry are now available. They permit important gains of operator time. Balancing the radiometry of images is a complex problem, difficult to model, in particular for scanned images (mostly due to the imperfections of the photographic acquisition, see §1.8). Problems of colour are due to movements of the sun and variations of the angle optical axis–sun. It generates problems of hot spot and of specular reflections. One zone has different aspects therefore according to the chosen image.

3.8.2 Line of join up

The automatic calculation of join-up lines on images of good radiometric quality is a relatively simple problem. Let's place us in the zone of the two orthoimages superposition. In this zone, every ground point possesses two representations: one in each orthoimage. The line of join-up is merely a path joining two opposite sides of this zone. One wants this line to be as invisible as possible. For that, one is going to assign a cost value to every point of the recovery zone. This value represents the 'visibility' of a join-up line if it passes by this point. A measure can be the difference between the grey level of pixels. Another measure can be the presence of a contrast between the two images. The line of join-up will then be the minimum cost path. (See Figure 3.8.2.)

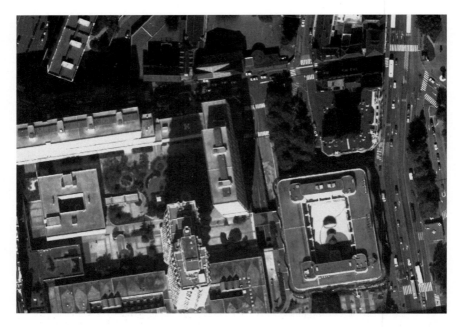

Figure 3.8.2 Automatic join-up example between four digital images.

3.8.3 Radiometric balancing

As is clearly visible on Figure 3.8.3 (colour section), radiometric differences between images, especially scanned images, can be very important. They may have many causes, such as the chemical aspects of the development, the adjustment of the scanner and the movements of the sun. Misleading differences between images can be very bothersome. Besides, these differences can put in failure the algorithm of calculation of join-up line presented above. For scanned images, the problem of modelling the difference is very complex, or even impossible. The used methods are therefore generally empirical. They use parameters that must be adjusted by hand and require a certain experience.

One of the essential problems is the volume of data. Thus, the regular cover (every 5 years) of French territory performed by IGN-F by a colour orthophotography of ground pixel 50 cm represents 1,200 Go of data. It is necessary to bring in at least as much data for images, then for overlap zones as well as auxiliary data. It represents 3 Tera-Bytes (4,800 CD-ROM) of data that must transit production lines and be archived every year. Material and software solutions exist, but are relatively heavy to operate. Obviously this part of the problem will progressively disappear with the evolution of informatics.

Production of digital orthophotographies 291

Figure 3.8.3 Scanned images before balancing.

Figure 3.8.4 Scanned images after balancing.

3.8.4 Conclusion

Having examined the bases of the orthophotography process, we will see in the next section (§3.9) how to perform the practical production itself. An important point must nevertheless be raised now: as mentioned in §3.7 the production of orthophotographies is quite different if we are in urban or in rural zones. In rural areas, the main quality aspect will come from the correct mosaicking (particularly the problem of 'hot spot'), nobody will notice the way the trees or the fences are represented. In urban areas, the representation of the buildings induces another set of difficulties

292 Didier Boldo

already mentioned (hidden zones, very dark shadows, etc.). Thus the software used for such zones are mostly different. This must be taken into account when one thinks of a production line.

3.9 PROBLEMS RELATING TO ORTHOPHOTOGRAPHY PRODUCTION

Didier Moisset

The orthophotography is an *artificial* digital picture. Completely manufactured by the computer, it looks automatically in photographs for radiometry information and borrows from the map, of automatic way, its carroyage, its bootjacks and sometimes some vector complements (contour lines, roads, site names, etc.).

It is neither a photo nor a map and the customer wants it to be both as beautiful as the photo and as precise as the map. This ambiguousness can cause some problems for the producer of orthophotos. One used to say that the data size was a major problem. The advent of the fast networks and very high capacity drives allied to the increasing strength of machines makes this affirmation less and less true. Since the production is fully computerized, it is legitimate to want it completely automatic and this is the case most of the time when the image acquisition is good, the landscape simple, and the control data of good quality. One will notice that these conditions are sometimes present (particularly on the data sets that are used for software demonstrations!), but only 'sometimes'.

In a situation of mass production, conditions are sometimes very different and problems occur quickly when, in spite of the power of our machines, data don't permit, by their nature, the association of the aesthetic quality

Figure 3.9.1 Example from a real production (digital aerial camera).

Problems of orthophotography production 293

of the photo and the precision of the map. Of course, it is when he counts in its significant cost that the volume of data manipulated becomes appreciable for the producer.

The homogeneity of the radiometry

Let's imagine an automatic production chain that would only put in evidence at the time of the mosaicking that the nature of the digital pictures originating from the scanner do not allow a correct radiometric equating to be performed. (See Figure 3.9.2, colour section.)

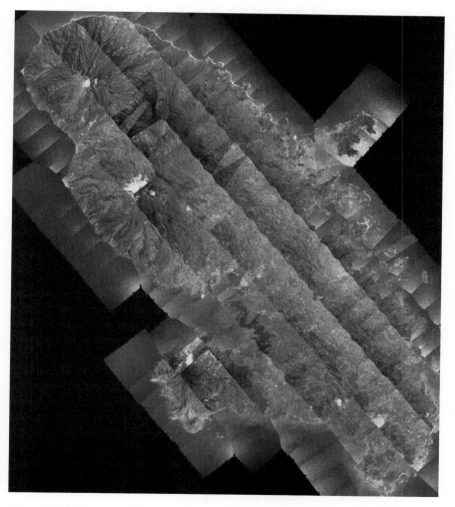

Figure 3.9.2 Example from a real production problem: the radiometric balancing on digitized pictures.

This example from a real production is a caricature that shows us the importance of an efficient procedure of validation of scanning operations for the production chain. The replacement of a film during the aerial mission, the modification of chemical film process or a problem in the calibration of the scanner are many factors that can occur and generate locally for the mosaic bothersome shortcomings as those noted here.

These problems don't exist as soon as one uses digital cameras, but let us keep in mind that when a mission spreads in time, the modification of the landscape can become appreciable, or even intolerable for the mosaic.

The interpretation of the banking of the raised structures

By its very nature, the orthophoto even when it is of high quality is able to remind us that it is not a photo and sometimes the choice in the position of the join-up line has a fundamental importance for the aesthetics of the mosaic (see Figures 3.9.3 (a) and (b)).

The producer must prevent this kind of fantasy and must use possibilities of modification of the automatic calculation of the join-up line by imposing points of passage if its tool possesses this functionality. If it is not the case, he should modify the join-up line by hand making it follow the characteristic lines of the landscape.

This example is also caricatural, but one can imagine without difficulty the problem for the producer of the orthophoto who must define and value the level of sharpness that he needs to grant for the verification and the possible modification of join-up lines. It is therefore a costly interactive phase that is going to condition the quality of the final result.

Stretch of pixels

Another difficulty that one frequently meets on landscapes to strong reliefs: when the optical ray is tangent to the DTM, a displacement on the DTM doesn't generate a displacement on the image. The image will provide the same radiometry therefore for a succession of pixels of the orthophoto. One says that the pixel is *stretched*. (See Figures 3.9.4 and 3.9.5, colour section.)

In this case, which is not rare, the choice of the image that will provide the radiometry (therefore the position of the join-up line) is determinant for the quality of the resulting mosaic.

Difficulties bound to the type of landscape

Some landscapes create some difficult problems to solve. The following are two examples of landscapes whose aspect changes significantly according to the point of view from which one looks at it.

Figures 3.9.3 (a) and (b) Example from a real production (area of Dijon). A correct line of join-up is not just a question of radiometry, especially in urban zones.

Figure 3.9.4 Examples from the Ariège, France: problems posed by cliffs.

Figure 3.9.5 Generation of stretched pixels.

Sometimes, the join-up line is difficult to conceal whereas there is no alternative. It is the case in landscapes of strong reliefs where the aspect of vegetation is going to depend greatly on the angle from which one looks at it (see Figure 3.9.6, colour section).

It is also the case with specular reflections well known from picture processors (Figure 3.9.7, colour section).

Figure 3.9.6 Example of radiometric difficulties (Ariège, France).

Figure 3.9.7 Example of typical problems on water surfaces.

Figure 3.9.8 Two examples from the Isère department, France.

Difficulties bound to the geometry

On the examples shown in Figure 3.9.8 (colour section), the radiometric homogenization (that is acceptable) is going to overlook for a non-aware eye a serious defect of geometry. It is clear in this case that the DTM and the result of the aerotriangulation that have both served to elaborate the orthophotography are not in agreement.

When this observation is performed only via the examination of the mosaic (which is the final product) one easily understands the consequences for the producer who must analyse the whole chain starting with the control points of the aerotriangulation, as well as the nature of difficulties that is going to meet the person who must value the amount of work to perform at this step.

But when all goes well

The examples shown in Figure 3.9.9 (colour section) allow one to conclude on a positive note. When geometry and radiometry are in harmony, and in particular when the radiometry is very precise (the case of the digital cameras acquisitions), it is necessary to make the join-up line visible in the mosaic to remind one that the orthophotography is not originating from only one unique photo.

By these examples, one can understand the importance that the producer of orthophotographies will pay to all operations of validation of the different steps of the process.

As it uses a set of high technologies ranging from photogrammetry to the development of digital models of land and digital picture processes, the rigorous realization of an orthophotography at a time is not a simple thing, even if it presents the strong interest of being automated.

Problems of orthophotography production 299

Figure 3.9.9 Two examples, using the digital aerial camera of IGN-F. Due to the linearity of the CCD, there are only very minor radiometric differences across the join-up line, which may be chosen without difficulty.

4 Metrologic applications of digital photogrammetry

INTRODUCTION

At the end of this book, we have a short presentation of some specific applications of digital photogrammetry, generally at closer ranges and without aerial acquisition of the images. Two main domains are presented here, the architectural applications (§4.1) and the metrologic applications (§4.2). These two domains have experienced a considerable extension in recent years, mainly because of the use of digital photogrammetry and the availability of low-cost digital photogrammetric workstations (DPW).

4.1 ARCHITECTURAL PHOTOGRAMMETRY

Pierre Grussenmeyer, Klaus Hanke, André Streilein

4.1.1 Introduction

Compared with aerial photogrammetry, close-range photogrammetry and particularly architectural photogrammetry isn't limited to vertical photographs with special cameras. The methodology of terrestrial photogrammetry has changed significantly and various photographic acquisitions are widely in use.

New technologies and techniques for data acquisition (CCD cameras, Photo-CD, photoscanners), data processing (computer vision), structuring and representation (CAD, simulation, animation, visualization) and archiving, retrieval and analysis (spatial information systems) are leading to novel systems, processing methods and results.

The purpose of this chapter is to introduce, as part of the International Committee for Architectural Photogrammetry (CIPA), the profound changes currently stated.

The improvement of methods for surveying historical monuments and sites is an important contribution to the recording and perceptual monitoring of cultural heritage, to the preservation and restoration of any

Architectural photogrammetry

valuable architectural or other cultural monument, object or site, as a support to architectural, archaeological and other art-historical research.

4.1.2 Strategies for image processing

Close-range photogrammetry is a technique for obtaining geometric information, e.g. position, size and shape of any object, that was imaged on photos previously.

To achieve a restitution of a 3D point you need the intersection between at least two rays (from photo to object point) in space or between one ray and the surface that includes this point. If more than two rays are available (the objects shows on three or more photos) a bundle solution is possible including all available measurements (on photos or even others) at the same time.

These cases lead to different approaches for the photogrammetric restitution of an object.

4.1.2.1 Single images

A very common problem is that we know the shape and attitude of an object's surface in space (digital surface model) but we are interested in the details on this surface (patterns, texture, additional points, etc.). In this case a single image restitution can be appropriate.

With known camera parameters and exterior orientation

In this case the interior orientation of the camera and the camera's position and orientation are needed. So the points can be calculated by intersection of rays from camera to surface with the surface known for its shape and attitude.

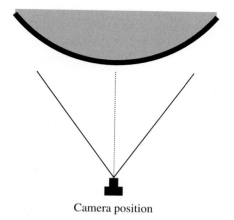

Figure 4.1.1

Interior orientation does not mean only the calibrated focal length and the position of the principal point but also the coefficients of a polynomial to describe lens distortion (if the photo does not originate from a metric camera).

If the camera position and orientation is unknown at least three control points on the object (points with known coordinates) are necessary to compute the exterior orientation (spatial resection of camera position).

Without knowledge of camera parameters

This is a very frequent problem in architectural photogrammetry. The shape of the surface is restricted to planes only and a minimum number of four control points in two dimensions have to be available. The relation of the object plane to the image plane is described by the projective equation of two planes:

$$X = \frac{a_1 x + a_2 y + a_3}{c_1 x + c_2 y + 1},$$

$$Y = \frac{b_1 x + b_2 y + b_3}{c_1 x + c_2 y + 1},$$

where X and Y are the coordinates on the object's plane, x and y the measured coordinates on the image and a_i, b_i, c_i the eight parameters describing this projective relation.

The measurement of a minimum of four control points in the single photo leads to the evaluation of these eight unknowns ($a_1, a_2, a_3, \ldots, c_2$).

As a result, the 2D coordinates of arbitrary points on this surface can be calculated using those equations. This is also true for digital images of facades. Digital image processing techniques can apply these equations for

Original image

Rectified orthophoto (in scale)

Figure 4.1.2

every single pixel and thus produce an orthographic view of the object's plane, a so-called orthophoto or orthoimage (see §3.7). (See Figure 4.1.2.)

4.1.2.2 *Stereographic processing*

If its geometry is completely unknown, a single image restitution of a 3D object is impossible. In this case the use of at least two images is necessary. According to the stereographic principle a pair of 'stereo images' can be viewed together, which produces a spatial (stereoscopic) impression of the object. This effect can be used to achieve a 3D restitution of, for example, facades. (See Figure 4.1.3.)

Using 'stereo pairs of images' arbitrary shapes of a 3D geometry can be reconstructed as long as the area of interest is shown on both images. The camera directions should be almost parallel to each other to have a good stereoscopic viewing.

Metric cameras with well known and calibrated interior orientation and negligible lens distortion are commonly used in this approach. To guarantee good results the ratio of stereo base (distance between camera positions) to the camera distance to the object should lie between 1:5 and 1:15.

Results of stereographic restitution can be:

- 2D-plans of single facades;
- 3D-wireframe and surface models;
- lists of coordinates;
- eventually complemented by their topology (lines, surfaces, etc.).

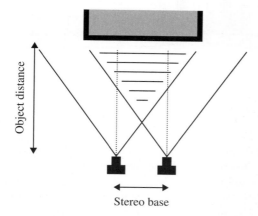

Figure 4.1.3

Figure 4.1.4 Stereopair from CIPA-Testfield 'Otto Wagner Pavillion Karlsplatz, Vienna'.

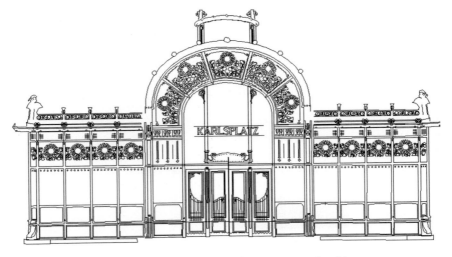

Figure 4.1.5 2D façade plan derived from above stereo pair of images.

4.1.2.3 Bundle restitution

In many cases the use of one single stereo pair will not suffice to reconstruct a complex building. Therefore a larger number of photos will be used to cover an object as a whole. To achieve a homogeneous solution for the entire building and also to contribute additional measurements, a simultaneous solution of all the photo's orientation is necessary.

Another advantage is the possibility to perform an on-the-job calibration of the camera. This helps to increase the accuracy when using images of an unknown or uncalibrated camera. So this approach is no longer restricted to metric or even calibrated cameras, which makes the applica-

Architectural photogrammetry 305

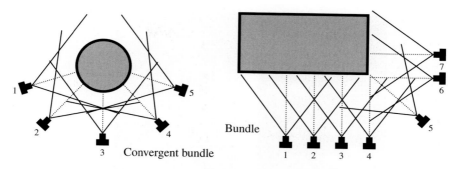

Figure 4.1.6 Examples of different configurations for bundle solution.

tion of photogrammetric techniques much more flexible. It is also adjustable concerning the geometry of camera positions, meaning one is not forced to look for parallel views and stereo pair configuration. Convergent, horizontally, vertically or obliquely photos are now certainly suitable. Combination of different cameras or lenses can easily be done.

The strategy of taking photos is that each point to be determined should be intersected by at least two rays of satisfactory intersection angle. This angle depends only upon the accuracy requirements. Additional knowledge of, for example, parallelism of lines, flatness of surfaces and rectangularity of features in space can be introduced in this process and helps to build a robust and homogeneous solution for the geometry of the object.

The entire number of measurements and the full range of unknown parameters are computed within a statistical least squares adjustment. Due to the high redundancy of such a system it is also possible to detect blunders and gross errors, so not only accuracy but also reliability of the result will usually be increased.

Bundle adjustment is a widespread technique in the digital architectural photogrammetry of today. It combines the application of semi-metric or even non-metric (*amateur*) cameras, convergent photos and flexible measurements in a common computer environment. Because of the adjustment process, the results are more reliable and accurate and very often readily prepared for further use in CAD environments.

Results of bundle restitution are usually 3D-wireframe and surface models of the object or lists of coordinates of the measured points and their topology (lines, surfaces, etc.) for use in CAD and information systems. Visualizations and animations or so-called 'photo-models' (textured 3D-models) are also common results. Usually the entire object is reconstructed in one step and the texture for the surface is available from original photos (see §4.1.6).

Figure 4.1.7 Examples of different images, different cameras, different lenses (from project Ottoburg, Innsbruck) to combine within a bundle solution (Hanke and Ebrahim, 1999).

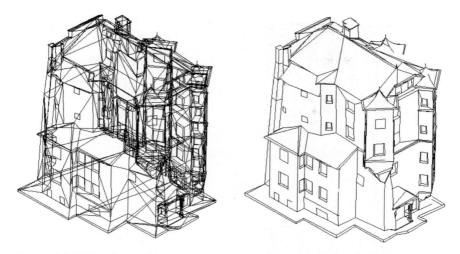

Figure 4.1.8 Wireframe model and surface model as results of bundle restitution.

4.1.3 Image acquisition systems

4.1.3.1 General remarks

Digital image data may be acquired directly by a digital sensor, such as a CCD array camera (see §1.5), for architectural photogrammetric work. Alternatively, it may be acquired originally from a photograph and then scanned (see §1.8).

For the applications in architectural photogrammetry the use of cameras was for a long time determined by the use of expensive and specialized equipment (i.e. metric cameras). Depending on the restrictions due to the photogrammetric reconstruction process in former times, only metric cameras with known and constant parameters of interior orientation could be used. Their images had to fulfil special cases for the image acquisition (e.g. stereo normal case).

Nowadays more and more image acquisition systems based on digital sensors are developed and available at reasonable prices on the market. The main advantage of these camera systems is the possibility to acquire digital images and directly process them on a computer.

Figure 4.1.9 gives a principal overview of the basic photogrammetric systems for image acquisition and image processing in architectural photogrammetry. The classic, photographic cameras have their advantages in the unsurpassed quality of the film material and resolution and in the well-known acquisition technique. The process of analytical photogrammetry makes a benefit of the knowledge and rich experiences of the human operator. On the other hand the pure digital data flow has not yet image

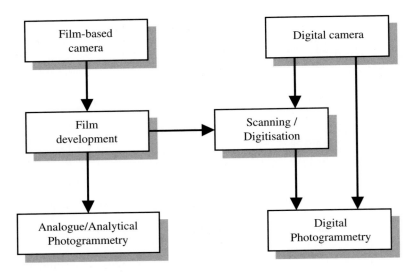

Figure 4.1.9 Image acquisition and image processing systems in architectural photogrammetry.

acquisition devices comparable to film-based cameras. But this procedure allows a productive processing of the data due to the potential of automation and the simultaneous use of images and graphics. Furthermore, it allows a closed and therefore fast and consistent flow of data from the image acquisition to the presentation of the results. In addition, with the digitization of film a solution is available that allows the benefits of the high-resolution film to be merged with the benefits of the digital image processing. But the additional use of time for the separate process of digitization and the loss of quality during the scanning process are disadvantages.

In the following sections the main photographic and digital image acquisition systems used in architectural photogrammetry are explained, examples are shown and criteria for their use are given.

4.1.3.2 *Photographic cameras*

From a photogrammetric point of view, film-based cameras can be subdivided into three main categories: metric cameras, stereo cameras and semi-metric cameras (see Table 4.1.1).

Terrestrial metric cameras are characterized by a consequent optical-mechanical realization of the interior orientation, which is stable over a longer period. The image coordinate system is, like in aerial metric cameras, realized by fiducial marks. In architectural photogrammetry such cameras

Table 4.1.1 Examples from the variety of film-based image acquisition systems

Manufacturer	Type	Image format [mm²]	Lenses [mm]
Metric cameras			
Hasselblad	MK70	60 × 60	60, 100
Wild	P32	65 × 90	64
Wild	P31	100 × 130	45, 100, 200
Zeiss	UMK 1318	130 × 180	64, 100, 200, 300
Stereo cameras			
Wild	C 40/120	65 × 90	64
Zeiss	SMK 40/120	90 × 120	60
Semi-metric cameras			
Rollei	3003	24 × 36	15–1000
Leica	R5	24 × 36	18–135
Rollei	6006	60 × 60	40–350
Hasselblad	IDAC	55 × 55	38, 60, 100
Pentax	PAMS 645	40 × 50	35–200
Linhof	Metrica 45	105 × 127	90, 150
Rollei	R_metrica	102 × 126	75, 150
Rollei	LFC	230 × 230	165, 210, 300
Geodetic Services	CRC-1	230 × 230	120, 240, 450

are less and less used. Amongst those cameras, which are still in practical use, are, for example, the metric cameras Wild P31, P32 and Zeiss UMK 1318. Such image acquisition systems ensure a high optical and geometrical quality, but are also associated with high prices for the cameras themselves. In addition they are quite demanding regarding the practical handling. Besides the single metric cameras in heritage documentation, often stereo cameras are used. These cameras are composed of two calibrated metric cameras, which are mounted on a fixed basis in standard normal case. A detailed overview of terrestrial metric cameras can be found in many photogrammetric textbooks (e.g. Kraus, 1993).

With the use of **semi-metric cameras** the employment of *réseau* techniques in photographic cameras was established in the everyday work of architectural photogrammetry. The *réseau*, a grid of calibrated reference marks projected on to the film at exposure, allows the mathematical compensation of film deformations, which occur during the process of image acquisition, developing and processing. Different manufacturers offer semi-metric cameras at very different film formats. Based on small and medium format SLR-cameras, systems exists from, for example, Rollei, Leica and Hasselblad. Their professional handling and the wide variety of lenses and accessories allow a fast and economic working on the spot. Semi-metric cameras with medium format offer a good compromise between a large-image format and established camera technique. An overview on semi-metric cameras is given in Wester-Ebbinghaus (1989).

Often in architectural applications so called **amateur cameras** are used. This is not for dedicated photogrammetric projects, but in emergency cases, where no other recording medium was available or in case of destroyed or damaged buildings when only such imagery is available. Examples are given in, amongst others, Gruen (1976), Dallas *et al.* (1995) and Ioannidis *et al.* (1996). Due to the ongoing destruction of the world cultural heritage it will also be necessary in the future to reconstruct objects taken with amateur cameras (Waldheusl and Brunner, 1988).

4.1.3.3 Scanners

The digitization of photographic images offers a means to combine the advantages of film-based image acquisition (large image format, geometric and radiometric quality, established camera technique) with the advantages of digital image processing (archiving, semi-automatic and automatic measurement techniques, combination of raster and vector data).

Scanners for the digitization of film material can be distinguished regarding different criteria. For example, regarding the type of sensor, either point, line or area sensor, or regarding the arrangement with respect to the scanned object as flatbed or drum scanner (see §1.8).

For the practical use of scanners in architectural photogrammetric applications the problem of necessary and adequate scan resolution has to be faced. On the one hand the recognition of details has to be ensured and on the other hand the storage medium is not unlimited. This holds especially for larger projects. To scan a photographic film with a resolution equivalent to the film a scan resolution of about 12 μm (2,100 dpi) is required. Thus, a scanned image from a medium format film (6×6 cm^2) has about $5,000 \times 5,000$ pixels. To hold this data on disk requires approximately 25 Mbytes for a black-and-white scan and 75 Mbytes for a coloured image. For a scanned colour aerial image one would get a digital image of $20,000 \times 20,000$ pixels or 1.2 Gbytes. Even with the constant increasing size and decreasing costs for computer storage medium, this is a not to underestimate this factor in the planning of a project.

For the use in architectural photogrammetry typically two different types of scanners are used, high-resolution photogrammetric scanners and desktop publishing scanners.

The **photogrammetric scanners** are typically flatbed scanners, which have a high geometric resolution (5–12.5 μm) and a high geometric accuracy (2–5 μm). Currently there are just a few systems commercially available, which are offered mainly by photogrammetric companies. An overview on existing systems is given in Baltsavias and Bill (1994).

The **desktop publishing scanners** (DTP) are not developed for photogrammetric use, but they are widely available on the market at low cost and they are developed and improved in a short time interval. DTP scanners have typically a scan size capability of A4 or A3 formats with a scan resolution of 300–1,200 dpi. The geometric resolution of these systems is about 50 μm. Despite this technical reduction compared to photogrammetric scanners, these scanners, which are low cost and easy to handle, can be used for photogrammetric purposes. This holds especially for calibrated systems, where geometric accuracy in the order of 5–10 μm is feasible (Baltsavias and Waegli, 1996).

Another possibility for the digitization and storage of film material is offered by the **Photo-CD system**. Small and medium format film can be digitized in a special laboratory and stored on CD-ROM. The advantage of such a system is the inexpensive and easy digitization and convenient data archiving. On the other hand the scanning process cannot be controlled or influenced and the image corners are usually not scanned. Thus the interior orientation of an image is nearly impossible to reconstruct. Investigations about the practical use of the Photo-CD system for digital photogrammetry are performed by Hanke (1994) and Thomas *et al.* (1995).

4.1.3.4 CCD cameras

The development of digital image acquisition systems is closely connected to the development of CCD sensors. The direct acquisition of digital images with a CCD sensor holds a number of advantages, which makes them interesting for photogrammetric applications. For example:

- direct data flow with the potential of online processing;
- high potential for automation;
- good geometric characteristics;
- independent of the film development process;
- direct quality control of the acquired images;
- low-cost system components.

For photogrammetric applications mainly area-based CCD sensors are used. These sensors are produced for the commercial or industrial video market. Area-based CCD sensors are used in video cameras as well as in high resolution digital cameras for single exposures (still video cameras). Furthermore, there are specialized systems that use a scanning process for the image acquisition.

Standard **CCD video cameras** have a number of 400–580 × 500–800 sensor elements. With a pixel size of 7–20 μm these cameras have an image format between 4.4 × 6.0 mm and 6.6 × 8.8 mm. Such cameras deliver a standardized, analogue video signal with 25–30 images per second. This signal can be displayed on any video monitor. For the photogrammetric processing of this signal, it has to be digitized by a frame grabber.

Beside the entertainment industry such video cameras are mainly used for simple measurement or supervision tasks. The variety of systems on the market is enormous and can hardly be followed. Such systems are of importance for the documentation of destroyed objects of the world cultural heritage. Under certain circumstances the reconstruction of a building from video imagery is possible (Streilein, 1995).

One of the newer developments of digital image acquisition devices are CCD video cameras with a digital output. These cameras offer an A/D conversion in the camera body and have storage capacity for one acquisition, so that it is possible to store the image. The digital image can be delivered to a computer, the transfer of the image data is practically free of noise.

Due to the restrictions regarding resolution of CCD video cameras and due to the ongoing improvements in the CCD market, today more and more **high-resolution digital cameras** are used. Such cameras can be described as a combination of a traditional small-format SLR camera with a high-resolution CCD sensor replacing the film. The digital image data is stored directly in the camera body. In the photogrammetric community very well-known representatives of this type of camera are distributed from Kodak/Nikon under the Name DCS 420 and 460. They offer resolutions of $1,524 \times 1,012$ pixel and $3,060 \times 2,036$ pixel respectively. In addition, various manufacturers (e.g. Kodak, Leaf, and Rollei) offer camera systems with a resolution of about $2,000 \times 2,000$ pixels. The main advantage of such systems is the fast and easy image acquisition. This is achieved due to the fact that image acquisition, A/D conversion, storage medium and power supply are combined in one camera body. This allows one to transfer the images immediately to a computer and to judge the quality of the acquired images or directly process them.

Scanning cameras improve the resolution of the final image by sequential scanning procedure. Two scanning principles are distinguished, micro scanning and macro scanning (Lenz and Lenz, 1993). Micro scanning cameras use the principle of moving a CCD area sensor in parts of the pixel size over the sensor. The final image will be calculated from the different single images taken after the movement. The format of the final image is the same as that of the single images. The resolution of the final image is increased in horizontal and vertical directions by the number of the single images taken in each direction (microscan factor). Macro scanning cameras on the other hand move a CCD area sensor over a large image format. The position of the single images taken is determined either optical-numerically or mechanically, such that the single images can be combined into one high-resolution image. Both methods have in common that they should be used when object and camera do not move during the period of image acquisition (typically several seconds). Furthermore, the illumination conditions should remain constant during that period. This is practically achievable only under laboratory conditions, so that such

Architectural photogrammetry 313

Table 4.1.2 Examples from the spectrum of digital image acquisition devices

Manufacturer	Type	Number of pixel (H×V)	Image format [mm²]
Pulnix	TM-560	582 × 500	8.8 × 6.6
Canon	ION RC 560	795 × 576	8.8 × 6.6
Sony	XC-77	768 × 593	8.8 × 6.6
JVC	GRE-77	728 × 568	6.4 × 4.8
Canon	EOS-DCS 3	1268 × 1012	20.5 × 16.5
Kodak	Megaplus-1.4	1320 × 1035	9.0 × 7.0
Agfa	ActionCam	1528 × 1148	16.5 × 12.4
Kodak	DCS 420	1524 × 1012	14.0 × 9.0
Kodak	Megaplus-4.2	2048 × 2048	18.4 × 18.4
Rollei	ChipPack	2048 × 2048	31.0 × 31.0
Kontron	ProgRes 3000	2994 × 2320	8.6 × 6.5
Kodak	DCS 460	3060 × 2036	30.0 × 20.0
Canon	EOS-DCS 1	3060 × 2036	27.6 × 18.4
Dicomed	BigShot	4096 × 4096	60.0 × 60.0
Kodak	Megaplus-16.8	4096 × 4096	36.8 × 36.8
Agfa	StudioCam	4500 × 3648	36.0 × 29.0
RJM	JenScan	4500 × 3400	8.8 × 6.6
Jenoptik	Eyelike	6144 × 6144	28.6 × 28.6
Rollei	RSC	7000 × 5500	55.0 × 55.0
Zeiss	UMK HighScan	15414 × 11040	166.0 × 120.0

systems have minor importance for applications in architectural photogrammetry.

For applications in architectural photogrammetry line-based sensors can also be used for the generation of high-resolution images. An example of this type of sensor used for architectural photogrammetry is given in Reulke and Scheele (1997). In this example several buildings are recorded using the WAOSS/WAAC CCD line scanner, which incorporates three line sensors of 5,184 pixels each.

In Table 4.1.2 a few examples from the great variety of digital image acquisition systems on the market, which are suitable for applications in digital architectural photogrammetry, are given. This compilation is naturally incomplete and is just mentioned to give an idea about the variety. A good overview on digital image acquisition systems is nowadays available on different Internet pages, e.g. Plug-In Systems (2000), PCWebopaedia (2000), or the Internet pages of the different manufacturers.

4.1.3.5 Which camera to use?

For the question which camera to use for a specific photogrammetric task for heritage documentation, there is basically no common answer or simple rule. Often a photogrammetric project is as complex as the object itself and more often the use of an image acquisition device is determined by the budget of the project (read: use a camera system that is already available at no cost).

However, for a successful photogrammetric project several aspects have to be taken into account. For example, the maximum time limit for the image acquisition on the spot and for the (photogrammetric) processing of the data afterwards. Further criteria can be: the need for colour images or black-and-white images, the requested accuracy of the final model, the smallest object detail that can be modelled, the minimum and maximum of images for the project, the mobility and flexibility of the image acquisition system or the integration into the entire production process. But after all the price of the image acquisition system and the possibilities for further processing of the image data remain as the major factors for selecting a specific image acquisition system for a specific project.

4.1.4 Overview of existing methods and systems for architectural photogrammetry

4.1.4.1 General remarks

Architectural photogrammetry and aerial photogrammetry don't have the same applications and requirements. Most of the commercial digital photogrammetric workstations (DPWs) are mainly dedicated to stereoscopic image measurement, aerial triangulation, the digital terrain model (DTM) and orthoimages production from aerial and vertical stereopair images. A survey of these DPWs is presented by Plugers (1999). In this section we consider systems and methods which are rather low cost comparing to those well-known products developed mainly for digital mapping. Software packages for architectural photogrammetry may use different image types, obtained directly by CCD cameras or by scanning small and medium format metric or non-metric slides (see §4.1.3). The quality of digital images directly influences the final result: the use of low-resolution digital cameras or low-priced scanners may be sufficient for digital 3D visual models but not for a metric documentation. The systems may be used by photogrammetrists as well as by architects or other specialists in historic monument conservation, and run on simple PC-systems that suffice for special tasks in architectural photogrammetry.

According to the specific needs in architectural documentation, the different kinds of systems are based either on digital image rectification, or on monoscopic multi-image measurement or on stereoscopic image measurement (Fellbaum, 1992). Software of such systems is advanced in such a way that mass restitution and modelling is possible, if digital images are provided in a well-arranged way. Only some software packages are mentioned in this paragraph, and the aim is not to make a survey of existing systems.

Architectural photogrammetry

To compare different systems, the following topics can be considered (CIPA, 1999):

- the handling of a system;
- the flow of data;
- the management of the project;
- the import and export of data (image formats, parameter of interior and exterior orientation, control information, CAD information);
- the interior orientation;
- the exterior orientation (one step or two steps);
- the reconstruction of the object;
- the derived results in terms of topology, consistency, accuracy and reliability;
- the amount of photogrammetric knowledge necessary to handle the system.

4.1.4.2 Recommendation for simple photogrammetric architectural survey

For simple photogrammetric documentation of architecture, simple rules that are to be observed for photography with non-metric cameras have been written, tested and published by Waldheusl and Ogleby (1994).

These so-called '3 × 3 rules' are structured in:

1 three geometrical rules:
   i   preparation of control information;
   ii multiple photographic all-around coverage;
   iii taking stereopartners for stereo-restitution.

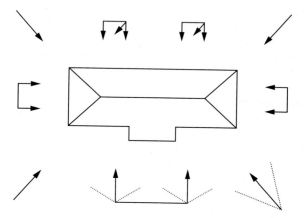

Figure 4.1.10 Ground plan of a stable bundle block arrangement all around building, as recommended in the 3 × 3 rules [see http://www.cipa.uibk.ac.at].

2   three photographic rules:
    i   the inner geometry of the camera has to be kept constant;
    ii   select homogeneous illumination;
    iii select most stable and largest format camera available.
3   three organizational rules:
    i   make proper sketches;
    ii   write proper protocols;
    iii don't forget the final check.

Usually, metric cameras are placed on a tripod, but shots with small or medium format equipment are often taken 'by hand'. Recently, digital phototheodolites combining total-station and digital cameras have been developed by Agnard *et al.* (1998), and Kasser (1998). Digital images are then referenced from object points or targets placed in the field. In this way, the determination of the exterior orientation is simple and the images are directly usable for restitution.

4.1.4.3 Digital image rectification

Many parts of architectural objects can be considered as plane. In this case, even if the photo is tilted with regard to the considered plane of the object, a unique perspective is enough to compute a rectified scaled image. We need at least four control points defined by their coordinates or distances in the object plane (§2.3.2).

The Rolleimetric MSR software package [http://www.rolleimetric.de] provides scale representations of existing objects on the basis of rectified digital images (Figures 4.1.11 (a), (b) and (c)). The base data is usually one or more photogrammetric images and/or amateur photographs of the object that are rectified at any planes defined by the user. Simple drawings (in vector-mode), image plans (in raster-mode) are processed as a result of the rectification.

Basically, the major stages encountered in the rectification of photography are as follows (Bryan *et al.*, 1999):

- site work (photography and control);
- scanning;
- rectification;
- mosaicking;
- retouching;
- output;
- archive storage.

Photographs of building *façades* should be taken the most perpendicular to the reference planes and only the central part of the image should be considered for a better accuracy.

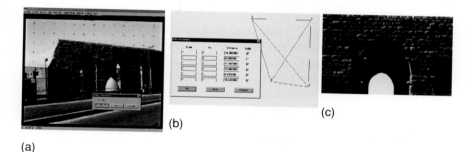

(a)

Figure 4.1.11 (a) Selection of a 'plane' part of the object. (b) Control distances for the plane definition. (c) Rectified image as an extract of (a).

Commercial CAD/CAM software packages often include image handling tools and also allow simple image transformation and rectification. But they seldom consider camera distortions, as opposed to photogrammetric software.

In the case of a perspective rectification, radial image displacements in the computed image will occur for points outside the reference. The rectification obviously fails if the object isn't somewhat plane.

Some packages include functions for the photogrammetric determination of planes according to the multi-image process (§4.1.4.4) from two or three photographs that capture an object range from different viewpoints. Digital image maps can be produced by assuming the object surface and photo rectification. In the resulting orthophoto, the object model is represented by a digital terrain model (Pomaska, 1998). Image data of different planes can be combined into digital 3D computer models for visualization and animation with the help of photo editing or CAD software.

ELSP from PMS [http://www.pms.co.at] and Photoplan [http://www.photoplan.net] are other examples of commercial systems particularly dedicated to rectification.

Besides digital image rectification and orthoimaging, single images for 3D surfaces of known analytical expression may lead to products in raster form, based on cartographic projections. Raster projections and developments of non-metric imagery of paintings on cylindrical arches of varying diameters and spherical surfaces are presented in Karras *et al.* (1997) and Egels (1998).

4.1.4.4 Monoscopic multi-image measurement systems

Photogrammetric multi-image systems are designed to handle two or more overlapping photographs taken from different angles of an object (see §4.1.2.3). In the past, these systems were used with analogue images enlarged and placed on digitizing tablets. Presently, the software usually processes image data from digital and analogue imaging sources (*réseau*, semi-metric or non-metric cameras). Scanners are used to digitize the analogue pictures. Film and lens distorsions are required to perform metric documentation. Monoscopic measurements are achieved separately on each image. These systems don't give the opportunity of conventional stereophotogrammetry. For the point measurements and acquisition of the object geometry, systems can propose support:

- for the automatic *réseau* cross-measurement;
- for the measurement of homologous points through the representation of line information in the photogrammetric images and epipolar geometry.

The acquisition of linear objects can be directly evaluated due to superimposition in the current photogrammetric image. These systems are designed for multi-image bundle triangulation (including generally simultaneous bundle adjustment with self-calibration of the used cameras, see §2.3). The main differences between the systems consist in the capabilities of the calculation module to combine additional parameters and knowledge, like directions, distances and surfaces. The main contribution of digital images in architectural photogrammetry is the handling of textures. The raster files are transformed into object surfaces and digital image data is projected on to a three-dimensional object model. Some systems are combined with a module of digital orthorectification.

The Canadian Photomodeler Software Package developed by Eos Systems is well known as a low-cost 3D-measurement tool for architectural and archeological applications [http://www.photomodeler.com]. Photomodeler (Figure 4.1.12) is a Windows-based software that allows measurements and transforms photographs into 3D models. The basic steps in a project performed with Photomodeler are:

- shoot two or more overlapping photographs from different angles of an object;
- scan the images into digital form and load them into Photomodeler;
- using the point and line tools, mark on the photographs the features you want in the final 3D model;
- reference the points by indicating which points on different photographs represent the same location on the object;
- process referenced data (and possibly the camera) to produce 3D model;
- view the resulting 3D model in the 3D viewer;
- extract coordinates, distances and areas measurements within Photomodeler;
- export the 3D model to rendering, animation or CAD program.

Architectural photogrammetry 319

Figure 4.1.12 Photomodeler's measuring module (project 'Ottoburg, Innsbruck', see §4.1.6).

Fast camera calibration based on a printable plane pattern can be set up separately to compute the camera's focal length, principal point, digitizing aspect ratio and lens distortion. Images from standard 35 mm film cameras digitized with Kodak Photo-CD, negative scanner or flatbed scanner as well as from digital and video cameras can be used in Photomodeler. The achieved accuracy obtained by Hanke and Ebrahim (1997) for distances between points lies in the range of 1:1,700 (for a 35 mm small format 'amateur' camera, without lens distortion compensation) to 1:8,000 (for a Wild P32 metric camera) and shows promising results.

Examples of systems based on the same concept are:

- KODAK Digital Science Dimension Software [http://www.kodak.com], with single and multiple image capabilities.
- 3D BUILDER PRO [http://aay.com/release.htm], with a 'constraint-based' modelling software package.
- SHAPECAPTURE [http://www.shapequest.com] offers target and feature extraction, target and feature 3D coordinate measurement, full camera calibration, stereo matching and 3D modelling.
- CANOMA [http://www.metacreations.com] from Meta Creations is a software intended for creating photorealistic 3D models from illustrations (historical materials, artwork, hand-drawn sketches, etc.), scanned or digital photographs. Based on an image-assisted technique, it enables one to attach 3D wireframe primitives and to render a 3D image by wrapping the 2D surfaces around these primitives.

Some other systems, also based on monoscopic multi-image measurement, are not mainly dedicated to the production of photomodels. In general, they use CAD models for the representation of the photogrammetric-generated results. For examples there are:

- CDW from Rolleimetric [http://www.rolleimetric.de] which is mainly a measuring system and doesn't handle textures. Data are exported to the user's CAAD system by interfaces. MSR^{3D}, also proposed by Rolleimetric, is an extension of MSR and is based on a CDW multi-image process (two or three photographs) for the determination of the different object-planes and the corresponding rectified images.
- Elcovision12 from PMS [http://www.pms.co.at] can run standalone or be directly interfaced with any CAAD application.
- PICTRAN from Technet (Berlin, Germany) [www.technet-gmbh.com] which includes bundle block adjustment and 3D restitution as well as rectification and digital orthophoto.
- PHIDIAS, proposed by PHOCAD (Aachen, Germany) is integrated in the Microstation CAD package.
- ORPHEUS (ORIENT-Photogrammetry Engineering Utilities System), proposed by the Institute of Photogrammetry and Remote Sensing of the Vienna University of Technology (Austria), is a digital measurement module running with the ORIENT software, linked with the SCOP software for the generation of digital orthophotos. It allows one, more particularly, to handle large images by the creation of image pyramids.

The prototype DIPAD (digital system for photogrammetry and architectural design) is based on the idea of using CAD models in a priori and a posteriori modes. Therefore CAD models are not only used for the graphical visualization of the derived results, but a complete integration between CAAD environment and photogrammetric measurement routines is realized. No manual measurements have to be performed during the whole analysis process. The system allows the three-dimensional reconstruction of objects or parts of it as an object-oriented approach in a CAAD-controlled environment. The task of object recognition and measurement is solved, with the image interpretation task performed by a human operator and the CAD-driven automatic measurement technique derives the precise geometry of the object from an arbitrary number of images. A human being uses intuitively, while looking at an image, his/her entire knowledge of the real world and is therefore able to select the information from or add missing information to the scene, which is necessary to solve the specific task. By choosing a combined top-down and bottom-up approach in feature extraction, a coarsely given CAD model of the object is iteratively refined until finally a detailed object description is generated. The semi-automatic HICOM method (human for interpretation and computer for measurement)

Figure 4.1.13 Screenshot of DIPA during the reconstruction process (interaction takes only place in the lower left CAD environment).

is able to simplify the analysis process in architectural photogrammetry without degradation in accuracy and reliability of the derived results. Typical problems for the automatic analysis of image data, which are very common in outdoor scenes, like occlusions of the object by itself or due to other objects (e.g. persons, vegetation), shadow edges and reflections, are detected with this method and attributed during further processing. Examples for the performance of this method are given in, for example, Streilein (1996) and (Streilein and Niederöst, 1998).

4.1.4.5 Stereoscopic image measurement systems

From analytical to digital

Digital stereoscopic measuring systems follow analytical stereoplotters well known as the more expensive systems. Many plottings are still done on analytical stereoplotters for metric documentation but as the performance and ability of digital systems increase and allow mass restitution. As textures are increasingly required for 3D models, digital photographs and systems are getting more and more important.

Stereoscopy

Systems presented in the former paragraph allow more than two images but homologous points are measured in monoscopic mode. Problems may occur for objects with less texture when no target is used to identify homologous points. Only stereo-viewing allows in this case a precise 3D measurement. Therefore stereopairs of images (close to the normal case) are required. Systems can then be assimilated to 3D plotters for the measuring of spatial object coordinates. 3D measurements are required for the definition of digital surface models that are the basis of the orthophotos. Usually, proposed solutions for stereo-viewing devices are:

- the split-screen configuration using a mirror stereoscope placed in front of the screen;
- the anaglyph process;
- the alternating of the two images on the full screen (which requires active glasses);
- the alternating generation of the two synchronized images on a polarized screen (which requires polarized spectacles).

Automation and correlation

In digital photogrammetry, most of the measurements can be done automatically by correlation. The task is then to find the position of a geometric figure (called the reference matrix) in a digital image. If the approximate position of the measured point is known in the image, then we can define a so-called search matrix. Correlation computations are used to determine the required position in the digital image. By correlation in the subpixel range the accuracy of positioning is roughly one order of magnitude better than the pixel size. Usually, the correlation process is very efficient on architectural objects, due to textured objects. Correlation functions can be implemented in the different steps of the orientation:

- fiducial marks or *réseau* crosses can be measured automatically in the inner orientation;
- measurement of homologous points can be automated by the use of the correlation process both in the exterior orientation, and in the digital surface model and stereoplotting modules.

The correlation function is a real step forward compared to manual measurements applied in analytical photogrammetry. The quality of the measurement is usually given by a correlation factor.

Model orientation

Systems currently used for aerial photogrammetry run with stereopairs of images but their use isn't always possible in architectural photogrammetry. In many cases, systems cannot handle different types and scales of images. The orientation module may fail either because photographs and control points are defined in a terrestrial case, or due to the non-respect of the approximate normal case. For the exterior orientation, some systems propose solutions based on typical two-step orientation (relative and absolute), as known in analytical photogrammetry. But bundle adjustment is increasingly proposed and allows an orientation in only one step. A direct linear transformation is sometimes used to compute approximate parameters of the orientation. Some systems propose automatic triangulation. But due to the configuration of the set of photographs in architectural photogrammetry, their use requires many manual measurements. After the relative orientation, often normalized images can be computed. Either original or normalized images can be used for further processes.

Stereo-digitizing and data collection

3D vectorization allows plottings (Figure 4.1.14) and digital surface model generation, with more or less semi-automatic procedures depending on the systems. Image superimposition is possible with almost every DPW. Some

Figure 4.1.14 Digital stereoplotting with image superimposition.

systems can be connected on-line to commercial CAD software packages and use their modelling procedures. Orthophotos can be generated from the different models but for architectural application, photomodels are usually carried out with multi-image systems (see Figures 4.1.6 and 4.1.7).

Examples of low-cost PC-systems based on stereoscopic measurements with known applications in close-range architectural photogrammetry, but only handling nearly normal case stereopairs, are:

- The digital video plotter (DVP) [http://www.dvp-gs.com]. It was one of the first systems proposed (Agnard et al., 1988). It is optimized for large-scale mapping in urban areas but architectural projects have been presented by DVP Geomatic Systems Inc.
- Photomod from Raccurs (Moscow, Russia) has a high degree of automatization, compared to other systems. The system is known as DPW for aerial photogrammetry. Orthophotos of facades with Photomod have been presented by Continental Hightech Services [http://www.chs-carto.fr].
- Imagestation SSK stereo Softcopy Kit from Intergraph Corporation is proposed as a kit to convert a PC into a low-cost DPW. Different modules of Intergraph ImageStation are available.

Several systems have been developed by universities during recent years. Some of them are:

- VSD as video stereo digitizer (Jachimski, 1995) is well known for architectural applications; VSD is a digital autograph built on the basis of a PC. It is suitable for plotting vector maps on the basis of pairs of monochrome or colour digital images as well as for vectorization of orthophotographs. The stereoscopic observation is based on a simple mirror stereoscope.
- POIVILLIERS 'E' developed by Yves Egels (IGN-ENSG, Paris, France) runs under DOS. Stereo-viewing is possible by active glasses connected to the parallel port of the PC or by anaglyphs. The system is very efficient for large images and high pointing accuracy is available due to a sub-pixel measuring module. Colour superimposition of plottings is also proposed. The system runs on aerial images as well as on terrestrial ones.
- Mapping from stereopairs within Autocad R14 is proposed by Greek Universities (Glykos et al., 1999).
- TIPHON is a Windows application developed at ENSAIS (Polytechnicum of Strasbourg, France) for two-image based photogrammetry (stereopair or convergent bundle) with different kinds of cameras (Grussenmeyer and Koehl, 1998). The measurements on the images are manual or semi-automatic by correlation. A stereoscope is used if stereoscopic observations are required.

- ARPENTEUR as 'ARchitectural PhotogrammEtry Network Tool for EdUcation and Research' is a platform-independent system available on the web by a simple internet browser (Drap and Grussenmeyer, 2000). The system is an adaptation of TIPHON to the Internet World and is particularly dedicated to architectural applications. ARPENTEUR is a web-based software package utilizing HTTP and FTP protocols. The photogrammetric adjustment and image processing routines are written in JAVA™. Different solutions are available for the orientation of the digital stereopairs. The concept of running photogrammetric software on the Internet is extended by a new approach of architectural photogrammetry 3D modelling. The architectural survey is monitored by underlying geometrical models stemming from architectural knowledge. Image correlation, geometrical functions and photogrammetry data are combined to optimize the modelling process. The data of the stereoplotting are directly superimposed on the images and exported towards CAD software packages and VRML file format. ARPENTEUR is usable via the Internet at [http://www.arpenteur.net].

4.1.5 3D object structures

4.1.5.1 General remarks

If a person is asked to describe an object, he/she solves the problem typically by describing all the single components of the object with all their attributes and properties and the relations they have with respect to each other and to the object. In principle computer representations and models are nothing other than the analogue description of the object, only the human language is replaced by mathematical methods. All kinds of representations describe only a restricted amount of attributes and each finite mathematical description of an object is incomplete.

Data models are necessary in order to process and manipulate real-world objects with the computer. The data models are abstractions of real-world objects or phenomena. Abstractions are used in order to grasp or manipulate the complex and extensive reality. Each attempt to represent reality is already an abstraction. The only complete representation of a real-world object is the object itself. Models are structures, which combine abstractions and operands into a unit useful for analysis and manipulation. Using models the behaviour, appearance and various functions of an object or building can be easily represented and manipulated. Prerequisite for the origin of a model is the existence of an abstraction. Each model needs to fulfil a number of conventions to work with it effectively. The higher the degree of abstraction the more conventions have to be fulfilled. CAAD models represent in an ideal way the building in form, behaviour and function as a logical and manipulable organism.

The data of the computer internal representation, which is sorted according to a specific order ('data structure'), forms the basis for software applications. The data basis is not directly accessed, but via available model algorithms, which allow the performance of complex functions by transforming them into simple basic functions according to a defined algorithm. The representation of a real-world object in a computer-oriented model is a synthesis of data structure and algorithms. Depending on extent and amount of information of the data, an object can be represented as a data-intensive or an algorithm-intensive model (Grätz, 1989). The most important role in the definition of models plays a proper balance between correctness and easy handling.

4.1.5.2 Classification of 3D models

In principle 3D models can be subdivided into three independent classes, the wireframe model, the surface model and the solid model (see Figure 4.1.15). The division is based on the different computer internal representation schemes and is therefore also for the application areas of these models.

Wireframe models are defined by the vertices and the edges connecting these vertices. They fix the outlines of an object and allow a view through from any point of view. This is an advantage for simple objects, but reduces the readability of more complex objects. This representation is therefore often used for simple objects.

Surface models represent the object as an ordered set of surfaces in three-dimensional space. Surface models are mainly used for the generation of models, whose surfaces consist of analytical not easily describable faces having different curvatures in different directions. This is often the case for models of automobiles, ships or aeroplanes.

Volumetric models represent three-dimensional objects by volumes. The data structure allows the use of Boolean operations as well as the calculation of volume, centre of gravity and surface area (Mäntylä, 1988). Surface modelling is the most demanding but also the most calculation-intensive way of modelling. Solid models always represent the hierarchy of the object, in which the primitives and operations are defined.

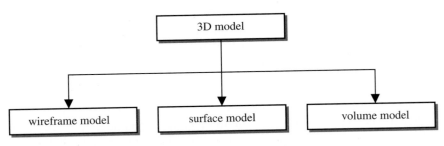

Figure 4.1.15 Overview of 3D models.

Each of the classes mentioned above has its specific advantages and disadvantages. Depending on the task the advantages and disadvantages are more or less important. Therefore it is not possible to make a general statement, which of the classes is the best representation of a real-world object.

Every representation of an object is a more or less exact approximation of the reality. A true-reality representation of a building considers all important attributes of the design. Looking at the section of a box object (see Figure 4.1.16) shows the differences of the different representations.

- The wireframe model represents the box as a quantity of vertices and edges. The section shows a quantity of non-connected points. This representation is reality-true if one is interested in the general form or in the position of the box.
- The surface model describes the box as a combination of vertices, edges and surfaces. The section shows a quantity of points and lines. This representation is reality-true if one is interested in the appearance of the surfaces.
- The volumetric model shows the box as a quantity of vertices, edges, faces and volume elements. The section shows points, lines and faces. This representation is reality-true if one is interested in mass properties, dynamic properties or material properties. For this purpose additional information that does not belong to the pure geometry of the object has to be evaluated.

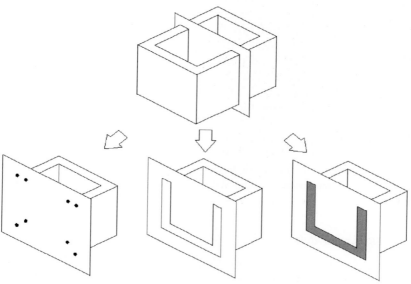

Figure 4.1.16 Reality-true representation of a box as wireframe, surface and volumetric model.

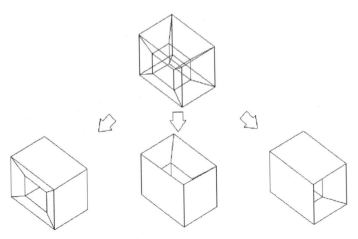

Figure 4.1.17 Example for ambiguity of a wireframe representation.

4.1.5.3 Wireframe models

The simplest forms of three-dimensional models are wireframe models. The computer internal model is confined basically on the edges of the object. Between those edges exists no relation, a relation to faces is not defined. Information about the inside and the outside of the object is not available. Points and edges are the only geometric elements of a wireframe model and are represented in the computer by list structures.

If a quantity of edges is stored with the condition that these edges should form the edges of a real-world object, it is often not unequivocal which object is represented. There exists more than one object that have the same edges as elements (see Figure 4.1.17).

The wireframe model allows therefore only an incomplete and ambiguous representation of a real-world object, which requires human interpretation. Manipulations are only possible on the basis of object edges (not object oriented) with the result that, as a result of geometric operations, nonsensical objects can be modelled. For example, deleting an edge of a cube results in a wireframe model which has no representation in the real world.

4.1.5.4 Surface models

The main information of the data structure of surface models is carried in the single surfaces of the model. These surfaces are typically not easy to describe by analytical means. A vector representation is also not possible. As a mathematical representation of such surfaces, approximate procedures

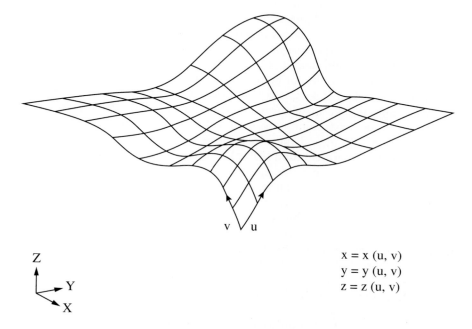

$$x = x(u, v)$$
$$y = y(u, v)$$
$$z = z(u, v)$$

Figure 4.1.18 Parametric representation of a surface (adapted from Grätz, 1989).

are often used, where the surface is represented between given points by a parametric function. The three-dimensional coordinates of the surface are a function of surface coordinates, which are used to parameterize the surface (see Figure 4.1.18).

Quite a number of mathematical procedures to handle surface models are available, such as Bézier approximations, spline and B-spline interpolations.

The surfaces in this data structure are quasi-independent elements, which means that they have no relations to each other. Surfaces do not share edges, the edges belong to the geometric description of the surfaces. A relation between a surface and a volume does not exist. Therefore models can be constructed, which are composed of a number of surfaces, but have no representation in the real world.

4.1.5.5 Volume models

The third class of 3D models are the volume models, which covers quite a number of different classes, such as parametric representation, sweep representation, octrees, cell decomposition, boundary representations, constructive solid geometry and hybrid models.

Characteristic for volume models is the computer internal volumetric representation of three-dimensional physical objects of the real word. These models are complete and non-ambiguous. Complete in this respect means that it is impossible to define or represent a volume with a missing surface or edge. On the other hand it is impossible to introduce surfaces or edges into the model, which do not belong to a volume. The produced computer model is in a technical sense a real representation of the reality. Therefore nonsense objects, like in wireframe and surface models, are impossible.

Amongst the most commonly used volume models are the boundary representation (BRep) and the constructive solid geometry (CSG).

The **boundary representation (BRep)** describes a three-dimensional object by its boundary elements (faces, edges and vertices). A boundary representation holds implicit a hierarchy in its data structure. An object is defined by its faces, the faces by its edges, the edges by its vertices and the vertices by three coordinates. A BRep model is organized in a topologic part and in a geometric part. Whereas the topologic part describes the neighbourhood relations between vertices, edges and faces, the geometric part holds the geometric information about the topologic elements (e.g. the three coordinates of the vertices). For the topologic relations have to obtain the following rules:

- each object is in relation to a number of faces, which bound it;
- each face is in relation to a number of edges, which bound it;
- each edge separates two faces from each other;
- each edge is in relation to a number of vertices, which limits it.

The topologic description of the object describes it as an arbitrary deformable object, which solidifies according to the geometric description.

Constructive solid geometry (CSG) describes the object instead of its bounding faces and edges by a set of defined volumetric primitives and basic operations of these primitives. The basic idea is that complex objects are represented as ordered additions, subtractions and intersections of simple objects. The primitives, which are used to sequentially build up the model, are directly visible in the data structure. Hence, basic volumes like cuboids, cones, globes, cylinders, etc. are contained in the data structure as primitives and directly accessible. They form the lowest level of geometric elements. An object is defined by the following grammar:

```
<object>           ::=  <primitive> |
                        <object> <transformation> argument |
                        <object> <operation> <object>
<primitive>        ::=  cube | cuboid | cylinder | cone | globe | ...
<transformation>   ::=  translation | rotation | scaling
<operation>        ::=  <combination> | <intersection> | <difference>.
```

Architectural photogrammetry

According to this grammar, each object is describable as a binary tree, where the leaves represent the primitives and the nodes represent the operations and transformations. The first node represents the modelled object. With this scheme the hierarchy of an object is implicitly given.

Figure 4.1.19 gives an example of the different data structures of boundary representation and constructive solid geometry of one and the same object.

4.1.5.6 Hybrid models

Beside the 'pure' representations of object models often hybrid models are utilized. Hereby the term 'hybrid model' is not clearly defined and is used for various types of models, which use different representations within one system. In principle the term can be used for all non-homogeneous models. It is often used for systems that can handle wireframe and surface models at the same time. Others use the term 'hybrid model' for using volume and surface models under the same graphical user interface. Or hybrid models are volume models, which use partly a CSG as well as a

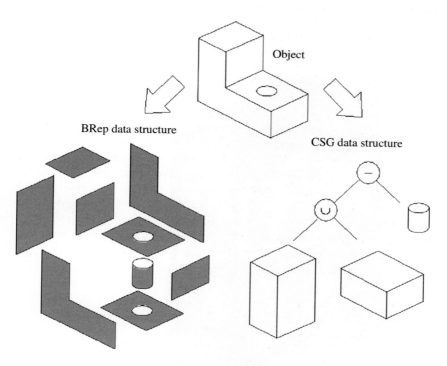

Figure 4.1.19 BRep and CSG data structure of the same object.

BRep data structure in order to have a dual representation of generative and accumulative elements. Hybrid models distinguish a primary structure, which is responsible for the exact model and all relevant model algorithms, and a secondary structure for specific tasks. The main task of this secondary structure is often the fast visualization of objects on the graphical screen.

4.1.6 Visual reality

Due to the progress in computer hard- and software there is a rapid development in the facilities of visualization in architectural photogrammetry. Simple facade plans are no longer suitable for the demands and applications of many users. 3D-real-time applications such as animations, interactive fly-overs and walk-arounds, which had needed the performance of high-end workstations a few years ago, are now also available on personal computers.

Two different concepts have to be distinguished. Where 'Virtual Reality' mainly uses vector models to describe a non-existing (= virtual) situation or fiction, 'Visual Reality' means a complex combination of vectors, surfaces and photo textures to visualize an existing object ('photomodel'). A fascinating idea is to merge these models into one, showing virtual objects within a visualized reality.

Figure 4.1.20 Perspective view of 3D photomodel 'Ottoburg, Innsbruck'.

Architectural photogrammetry 333

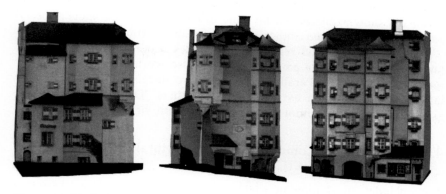

Figure 4.1.21 Orthoimages in scale of 3 facades of photomodel 'Ottoburg, Innsbruck'.

According to the purpose and required accuracy of the result there are numerous ways to create such textured 3D models. They range from sticking ortho-rectified photos to geometrically simplified surfaces of facades to a sophisticated reprojection of the original photos on to the complex geometry of a building using interior and exterior orientation of the camera.

Regarding these models of a monument's 3D data as a basic storage concept, a large number of resulting products can be derived from it. As examples arbitrary perspective views and orthoimages in scale should be referenced here.

A very promising way to visualize 3D-data is to create so-called 'worlds', not only for computer games but also for 'more serious' applications. VRML is a new standardized format (ISO 1997) describing three-dimensional models and scenes including static and dynamic multi-media elements. This description format is independent of the kind of computer. Most Internet browsers support VRML file format. 3D object models can be viewed and inspected interactively by the user or animated in real-time even on a PC. Thus, VRML is well suited to create, for example, interactive environments, virtual museums, visualizations and simulation based on real-world data.

Another way to visualize real-world objects is creating panoramic images. This approach avoids the time-consuming process needed for a 3D model. Plug-ins for Web browsers provide interactive movement. There are several methods to achieve panoramic images. One is to take single images with 20 per cent to 50 per cent overlap from a fixed position while rotating the camera around a vertical axis. Warping them on to a cylindrical or spherical surface leads to a spatial imagination when navigating through the model. Another way is to move the camera around the object with a fixed target point. Complex objects can be so viewed and turned around on a personal computer by simply dragging the mouse.

A combination of multiple panoramas or linking additional information about the shown objects can be done using clickable 'hot spots'. Special panoramic cameras and authoring tools for image stitching are available.

The new image-based rendering techniques using synthesis of images (among others to create panoramic views) also work without explicit 3D models of the object. Parallaxes between two images suffice for interpolation (sometimes even extrapolation) of others. They are restricted to existing objects and cannot be to combined with virtual worlds.

4.1.7 International Committee for Architectural Photogrammetry (CIPA)

To conclude this chapter in which an overview of recent developments and applications in architectural photogrammetry are given, we briefly present the International Committee for Architectural Photogrammetry (CIPA), as a forum on this field.

CIPA is one of the international committees of ICOMOS (International Council on Monuments and Sites) and it was established in collaboration with ISPRS (International Society of Photogrammetry and Remote Sensing).

Its main purpose is the improvement of all methods for the surveying of cultural monuments and sites, specially by synergy effects gained by the combination of methods under special consideration of photogrammetry with all its aspects, as an important contribution to the recording and perceptual monitoring of cultural heritage, to the preservation and restoration of any valuable architectural or other cultural monument, object or site, as a support to architectural, archaeological and other art-historical research.

ISPRS and ICOMOS created CIPA because they both believe that a monument can be restored and protected only when it has been fully measured and documented and when its development has been documented again and again, i.e. monitored, also with respect to its environment, and stored in proper heritage information and management systems.

In order to accomplish this mission, CIPA [see http://www.cipa.uibk.ac.at] will:

- establish links between architects, historians, archaeologists, conservationists, inventory experts and specialists in photogrammetry and remote sensing, spatial information systems, CAAD, computer graphics and other related fields;
- organize and encourage the dissemination and exchange of ideas, knowledge, experience and the results of research and development (CIPA Expert Groups and CIPA Mailing List);
- establish contacts with and between the relevant institutions and companies which specialize in the execution of photogrammetric surveys or in the manufacture of appropriate systems and instruments (Board of Sustaining Members);

- initiate and organize conferences, symposia, specialized colloquia, workshops, tutorials, practical sessions and specialized courses (CIPA Events);
- initiate and coordinate applied research and development activities (CIPA Working Groups);
- undertake the role of scientific and technical expert for specific projects (CIPA Expert Advisory Board);
- organize a network of National and Committee Delegates;
- submit an annual report on its activities to the ICOMOS Bureau (Secretary General) and the ISPRS Council (Secretary General) and publish it on the Internet (Annual Reports);
- publish also its Structure, its Statutes and Guidelines on the Internet.

CIPA has a well-established structure of Working Groups (WG) and Task Groups (TG):

- WG 1 – Recording, Documentation and Information Management Principles and Practices;
- WG 2 – Cultural Heritage Information Systems;
- WG 3 – Simple Methods for Architectural Photogrammetry;
- WG 4 – Digital Image Processing;
- WG 5 – Archaeology and Photogrammetry;
- WG 6 – Surveying Methods for Heritage Recorders;
- WG 7 – Photography;
- WG 8 – Cultural Landscapes;
- TG 1 – Non-professional Heritage Recorders;
- TG 2 – Single Images in Conservation.

Bibliography

References from books

Atkinson, K.B., 1996. *Close Range Photogrammetry and Machine Vision*. Whittles Publishing, London.

Batic J., et al., 1996. *Photogrammetry as a Method of Documenting the Cultural Heritage* (in English and Slovenian). Ministry of Culture, Ljubljana, Slovenia.

Dallas, R.W.A., 1996. Architectural and archaeological photogrammetry. In K.B. Atkinson (ed.) *Close Range Photogrammetry and Machine Vision*. Whittles Publishing, Caithness, UK, pp. 283–302.

Fondelli, M., 1992. *Trattato di fotogrammetria urbana e architettonica* (in Italian). Gius Laterza and Figli Spa (eds), Roma-Bari, Italy.

Grätz, J.F., 1989. *Handbuch der 3D-CAD-Technik. Modellierung mit 3D-Volumensystemen* (in German). Siemens-Aktiengesellschaft, Berlin.

Kraus, K. with contributions by Waldheusl P., 1993. *Photogrammetry, vol. 1, Fundamentals and Standard Processes*, 4th edition, Dümmler, Bonn, ISBN 3-427-78684-6. Translated into French by P. Grussenmeyer and O. Reis 1998

Manuel de photogrammétrie – principes et procédés fondamentaux, Editions HERMES, Paris.

Kraus, K. with contributions by J. Jansa and H. Kager, 1997. *Photogrammetry – Advanced Methods and Applications*. vol. 2, 4th edition, Dümmler, Bonn.

Luhmann, Th., 2000. *Nahbereichsphotogrammetrie – Grundlagen, Methoden und Anwendungen* (in German). Wichmann-Verlag, Heidelberg.

Mäntylä, M., 1988. *Introduction to Solid Modelling*. Computer Science Press, Maryland, US.

Patlas, P., Karras, G.E., 1995. *Contemporary Photogrammetric Applications in Architecture and Archaeology* (in Greek). University of Thessaloniki.

Saint-Aubin, J.-P., 1992. *Le relevé et la représentation de l'architecture* (in French). Inventaire Général des Monuments et des Richesses Artistiques de la France, Paris, 232pp. (in French).

Weimann, G., 1988. *Architektur-Photogrammetrie* (in German). Wichmann Verlag, Karlsruhe, Germany.

References from journals and other literature

Agnard, J.-P., Gagnon, P.-A., Nolette, C., 1988. Microcomputers and photogrammetry. A new tool. The Videoplotter. *PEandRS*, 54 (8), pp. 1165–1167.

Agnard, J.P., Gravel, C., Gagnon, P.-A., 1998. Realization of a digital phototheodolite. *International Archives of Photogrammetry and Remote Sensing*, vol. XXXII, Part 5, Hakodate, pp. 498–501.

Almagro, A., 1999. Photogrammetry for everybody. *International Archives of Photogrammetry and Remote Sensing*, vol. XXXII, CIPA Symposium, Olinda, Brazil.

Baltsavias, E., Bill, R., 1994. Scanners – a survey of current technology and future needs. *International Archives of Photogrammetry and Remote Sensing*, vol. 30, Part 1, pp. 130–143.

Baltsavias, E., Waegli, B., 1996. Quality analysis and calibration of DTP scanners. *International Archives of Photogrammetry and Remote Sensing*, vol. 31, Part B1, pp. 13–19.

Bryan, P.G., Corner, I., Stevens, D., 1999. Digital rectification techniques for architectural and archeological presentation. *Photogrammetric Record*, 16(93) (April), pp. 399–415.

Chengshuang, L., Rodehorst, V. Wiedemann, A., 1997. Digital image processing for automation in architectural photogrammetry. In O. Altan and L. Gründig (eds) *Second Turkish-German Joint Geodetic Days. Berlin, Germany, May 28–30, 1997*. Istanbul Technical University, pp. 541–548.

CIPA, 1999. Questionnaire on the processing of the data set 'Zurich city hall'. Edited by CIPA Working Group 3 and 4 (A. Streilein, P. Grussenmeyer and K. Hanke), 1999. 8 pp. Available at http://www.cipa.uikk.ac.at

Dallas, R.W.A., Kerr, J.B., Lunnon, S., Bryan, P.G., 1995. Windsor Castle: photogrammetric and archaelogical recording after the fire. *Photogrammetric Record*, 15(86), pp. 225–240.

Drap, P., Grussenmeyer, P., 2000. A digital photogrammetric workstation on the WEB. *ISPRS Journal of Photogrammetry and Remote Sensing*, 55(1), pp. 48–58.

Egels, Y., 1998. Monuments historiques et levers photogrammétriques. *Revue Géomètre*, (3), pp. 41–43, in French.

El-Hakim, S., 2000. A practical approach to creating precise and detailed 3D models from single and multiple views. *International Archives of Photogrammetry and Remote Sensing*, vol. 33, Part 5, pp. 203–210.

Fellbaum, M., 1992. Low-cost systems in architectural photogrammetry. *International Archives of Photogrammetry and Remote Sensing*, vol. XXIX, Part B5, Washington DC, pp. 771–777.

Glykos, T., Karras, G.E., Voulgaridis, G., 1999. Close-range photogrammetry within a commercial CAD package. *International Archives of Photogrammetry and Remote Sensing*, vol. XXXII, Part 5W11, Thessaloniki, pp. 103–106.

Gruen, A., 1976. Photogrammetrische Rekonstruktion aus Amateuraufnahmen. *Architektur-Photogrammetrie II, Arbeitsheft 17, Landeskonservator Rheinland.* Rheinland-Verlag, Köln, pp. 85–92.

Gruen, A., 2000. Semi-automated approaches to site recording and modelling. *International Archives of Photogrammetry and Remote Sensing*, vol. 33 Part 5, pp. 309–318.

Grussenmeyer, P., Abdallah, T., 1997. The cooperation in architectural photogrammetry between ENSAIS (France) and ECAE (Egypt): practical experiences on historic monuments in CAIRO. *International Archives of Photogrammetry and Remote Sensing*, vol. XXXII, Part 5C1B, Göteborg, pp. 215–221.

Grussenmeyer, P., Guntz, C., 1997. Photogrammétrie architecturale et réalité virtuelle: modélisation de l'aqueduc El-Ghuri (Le Caire, Egypte). *Revue de l'Association Française de Topographie*, XYZ 4e trim. 97 no. 73, pp. 75–81, in French.

Grussenmeyer, P., Koehl, M., 1998. Architectural photogrammetry with the TIPHON software, towards digital documentation in the field. *International Archives of Photogrammetry and Remote Sensing*, vol. XXXII, Part 5, Hakodate, pp. 549–556.

Hanke, K., 1994. The Photo-CD – A Source and Digital Memory for Photogrammetric Images. *International Archives of Photogrammetry and Remote Sensing*. vol. XXX Part 5, Melbourne, 1994, pp. 144–149.

Hanke, K., 1998. Digital close-range photogrammetry using CAD and raytracing techniques. *International Archives of Photogrammetry and Remote Sensing*, vol. XXXII, Part 5, Hakodate, pp. 221–225.

Hanke, K., Ebrahim, M.A-B., 1997. A low cost 3D-measurement tool for architectural and archaeological applications. *International Archives of Photogrammetry and Remote Sensing*, vol. XXXI, Part 5C1B, CIPA Symposium, Göteborg 1997, pp. 113–120.

Hanke, K., Ebrahim, M.A-B., 1999. The 'digital projector': Raytracing as a tool for digital close-range photogrammetry. *ISPRS Journal of Photogrammetry and Remote Sensing*, 54(1), Elsevier Science B.V., Amsterdam, pp. 35–40.

Hemmleb, M., Wiedemann A., 1997. Digital Rectification and Generation of Orthoimages in Architectural Photogrammetry. *International Archives of Photogrammetry and Remote Sensing* vol. XXXI Part 5C1B, CIPA Symposium, Göteborg, pp. 261–267.

Ioannidis, C., Potsiou, C., Badekas, J., 1996. 3D detailed reconstruction of a demolished building by using old photographs. *International Archives of Photogrammetry and Remote Sensing*, vol. XXXI, Part B5, Vienna, pp. 16–21.

Jachimski, J., 1995. Video stereo digitizer. A small digital stereophotogrammetric working station for the needs of SIT and other application. Polish Academy of Sciences. The Krakow Section. *Proceedings of the Geodesy and Environmental Engineering Commission, Geodesy 38*, pp. 71–91.

Karras, G., Patlas, P., Petsa, E., Ketipis, K., 1997. Raster projection and development of curved surfaces. *International Archives of Photogrammetry and Remote Sensing*, vol. XXXII, Part 5C1B, Göteborg, pp. 179–185.

Kasser, M., 1998. Développement d'un photothéodolite pour les levés archéologiques. *Revue Géomètre*, (3), pp. 44–45, in French.

Lenz, R., Lenz, U., 1993. New development in high resolution image acquisition with CCD-array sensor. In A. Gruen and H. Kahmen (eds) *Optical 3-D Measurement Techniques II*. Wichmann-Verlag, Karlsruhe, 1993. S. 53–62.

Patias, P., Peipe, J., 2000. Photogrammetry and CAD/CAM in culture and industry, an ever-changing paradigm. *International Archives of Photogrammetry and Remote Sensing*, vol. 33 Part 5, pp. 599–603.

Patias, P., Streilein, A., 1996. Contribution of videogrammetry to the architectural restitution – Results of the CIPA 'O. Wagner Pavillon' test. *International Archives of Photogrammetry and Remote Sensing*, vol. XXI, Part B5, Vienna, pp. 457–462.

PCWebopaedia, 2000. Digital camera. URL: www.pcwebopaedia.com/digital_camera.htm. (accessed November 2000).

Plugers, P., 1999. Product survey on digital photogrammetric workstations. *GIM International*, May, pp. 61–65.

Plug-In Systems, 2000. The digital camera guide. URL: www.pluginsystems.com. (accessed November 2000).

Pomaska, G., 1998. Automated processing of digital image data in architectural surveying. *International Archives of Photogrammetry and Remote Sensing*, vol. XXII, Part 5, Hakodate, pp. 637–642.

Reulke, R., Scheele, M., 1997. CCD-line digital imager for photogrammetry in architecture. *International Archives of Photogrammetry and Remote Sensing*, vol. XXXII, Part 5C1B, Göteburg, pp. 195–201.

Schneider, C-T., 1996. DPA-WIN – A PC-based digital photogrammetric station for fast and flexible on-site measurement. *International Archives of Photogrammetry and Remote Sensing*, vol. XXI, Part B5, Vienna, pp. 530–533.

Streilein, A., 1995. Videogrammetry and CAAD for architectural restitution of the Otto-Wagner-Pavilion in Vienna. In A. Gruen and H. Kahmen (eds) *Optical 3-D Measurement Techniques III*, Wichmann-Verlag, Heidelberg, pp. 305–314.

Streilein, A., 1996. Utilization of CAD models for the object oriented measurement of industrial and architectural objects. *International Archives of Photogrammetry and Remote Sensing*, vol. XXI, Part B5, Vienna, pp. 548–553.

Streilein, A., Beyer, H., Kersten, T., 1992. Digital photogrammetric techniques for architectural design. *International Archives of Photogrammetry and Remote Sensing*, vol. XXIX, Part B5, Washinton DC, pp. 825–831.

Streilein, A., Gaschen, S., 1994. Comparison of a S-VHS camcorder and a high-resolution CCD-camera for use in architectural photogrammetry. *International Archives of Photogrammetry and Remote Sensing*, vol. XXX, Part 5, Melbourne, pp. 382–389.

Streilein, A, Hanke, K., Grussenmeyer, P., 2000. First experiences with the 'Zurich City Hall' data set for architectural photogrammetry. *International Archives of Photogrammetry and Remote Sensing*, vol. 33 Part 5, pp. 772–779.

Streilein, A., Niederöst, M., 1998. Reconstruction of the Disentis monastery from high-resolution still video imagery with object-oriented measurement routines. *International Archives of Photogrammetry and Remote Sensing*, vol. XXII, Part 5, Hakodate, pp. 271–277.

Thomas, P.R., Miller, J.P., Newton, I., 1995. An investigation into the use of Kodak Photo-CD for digital photogrammetry. *Photogrammetric Record* 15(86), pp. 301–314.

Van den Heuvel, F.A., 1998, 3D reconstruction from a single image using geometric constraints. *ISPRS Journal of Photogrammetry and Remote Sensing*, 53(6), pp. 354–368.

Van den Heuvel, F.A., 1999. Estimation of interior orientation parameters from constraints on line measurements in a single image. *International Archives of Photogrammetry and Remote Sensing*, vol. 32, Part 5W11, Thessaloniki, pp. 81–88.

Varhosaz, M., Dowman, I., Chapman, D., 2000. Towards automatic reconstruction of visually realistic models of buildings. *International Archives of Photogrammetry and Remote Sensing*, vol. 33 Part 5, pp. 180–186.

Waldheusl, P., 1992. Defining the future of architectural photogrammetry. *International Archives of Photogrammetry and Remote Sensing*, vol. XXX, Part B5, Washington DC, pp. 767–770.

Waldheusl, P., 1999. Tasks for ISPRS working groups to serve ICOMOS. *International Archives of Photogrammetry and Remote Sensing*, vol. XXXII, Part 5W11, Thessaloniki, pp. 1–7.

Waldheusl, P., Brunner, M., 1988. Architectural photogrammetry world-wide and by anybody with non-metric cameras? *Proceedings des XI. Internationalen CIPA Symposiums*, October, Sofia, pp. 35–49.

Waldheusl, P.,Ogleby, C., 1994. 3 × 3-Rules for simple photogrammetric documentation of architecture. *International Archives of Photogrammetry and Remote Sensing*, vol. XXX, Part 5, Melbourne, pp. 426–429.

Wester-Ebbinghaus, W., 1989. Das Réseau im photogrammetrischen Bildraum. *Bildmessung und Luftbildwesen*, 3(89), pp. 64–71.

Wiedemann, A., 1996. Digital orthoimages in architectural photogrammetry using digital surface models. *International Archives of Photogrammetry and Remote Sensing*, vol. XXXI, Part B5, Vienna, pp. 605–609.

Wiedemann, A., Hemmleb, M., Albertz, J., 2000. Recontruction of historical buildings based on images from the Meydenbauer archives. *International Archives of Photogrammetry and Remote Sensing*, vol. 33 Part 5, pp. 887–893.

Wiedmann, A., Rodehorst, V., 1997. Towards automation in architectural photogrammetry using digital image processing. *International Archives of Photogrammetry and Remote Sensing*, vol. XXXI, Part 5C1B, CIPA Symposium, Göteborg, pp. 209–214.

Zischinsky, T., Dorffner, L., Rottensteiner, F., 2000. Application of a new model helicopter system in architectural photogrammetry. *International Archives of Photogrammetry and Remote Sensing*, vol. 33 Part 5, pp. 959–968.

4.2 PHOTOGRAMMETRIC METROLOGY

Michel Kasser

4.2.1 Introduction

The term 'photogrammetric metrology' covers the whole range of metrology activities that exploit photogrammetric processes, that is to say, geometric processes based on image acquisition, and the image processing which historically hardly ever took place in real time. The appearance of digital imagery has been the origin of a significant change in this technical domain, especially because it allowed real-time measures to be reached, which opened many new markets. And then the use of CCD cameras also permitted images of very large dynamics to be obtained, allowing the difficult cases formed by uniform surfaces, very current in industrial metrology, to be processed in a much better way. The very large dynamics makes it possible to detect very weak differences of radiometry having a meaningful character, which means they can be processed by automatic correlation.

Let us recall that photogrammetry is fundamentally a method founded on angle measures, angles that are recorded permanently on the film or in the digital pixel geometry. From this point of view it is easy to make the link with metrology methods based on theodolites measures, which are also used to measure angles, but with quite different devices. Therefore there is a continuity of methodologies between these two domains, and now that theodolites are digital, sometimes motorized, or capable of pointing their targets automatically, one notes without surprise that the same softwares are often used to process measures coming from these two techniques.

There is also a continuity between photogrammetry and studies in robotics to permit a fast identification of the geometric environment of the robot. The robotics developed some continual measures based on video images, allowing some elements in an image to be treated, this being performed at very high speed in order to allow the calculator to choose movements that the different motors have to perform. It is still photogrammetry, but a very specialized one, and that one sometimes calls videogrammetry.

Photogrammetric metrology finally represents all the methods using the equation of collinearity (see §1.1) on images, but in conditions that differ from the simple field of cartography. It represents a reuse of the body of knowledge of photogrammetry, but for applications that are each different, and that are rather specialized towards measures without contact with a large amounts of points. Objects processed are generally of restricted sizes, from about ten metres (wing of plane, cockles of boat) down to the millimetre, or even less (inspection of tapped holes, of soldering, of surfaces of samples checked with a microscope, etc.).

4.2.2 Equipment

The basis of modern digital photogrammetric metrology is the CCD camera. This type of sensor has been used with all available matrix sizes, from small-size matrixes for real-time applications up to 4k × 4k matrixes, and probably more sooner or later. This matrix is set up as usual in the focal plane of the optics, and as it is about doing measures these optics must have some steady geometric parameters in the time, so that they can make the object of a calibration. In this sense one doesn't search now, as in aerial imagery (see Chapter 2), for optics without distortion, but rather for optics having an excellent resolution, and whose distortion is steady in time, and the same for the position of their optic centre and their principal planes.

This general type (CCD + optics) can be used in very different ways. It can first of all be used locally, like a traditional camera, with successive positions of the same device or several synchronized devices, and while identifying the points in images on which are done the external geometric measures. The setting up of images is then achieved using traditional tools of photogrammetry, with possible adaptations for geometries very different from the aerial images with an almost vertical axis (e.g. problems of refraction).

It can also be used fixed very rigidly to the optic axis of a theodolite, which permits one to know the orientation in the reference frame corresponding to every image with the considerable precision of the system of angles measures of the theodolite. This whole is named phototheodolite; it was used a long time ago with film or even glass plates cameras, but has since fallen into obsolescence. If the theodolite is itself localized and oriented, which is a simple operation, one gets images that possess all parameters of localization and orientation, permitting an immediate setting up to do directly photogrammetric restitution. One can also use only the capacity of orientation of the theodolite to combine one set of images obtained in the different directions (motorized phototheodolite), and to get thus in a unique conical geometry an image having a very important number of pixels.

One can also try to measure rapidly a large number of targets (for only a few, a motorized theodolite with automatic aiming would be preferable), while going very quickly, typically in some milliseconds. In this case one uses retroreflecting targets, formed from many small plastic cubes' corners for example, and the optics used is then also equipped with an annular-shaped flash, which forms a circle completely surrounding the frontal lens. Thus targets send back a considerable light quantity towards the objective, and in return with a suitable threshold of the image one will see on this one only luminous points on a black scene, every point corresponding to a target. It will remain to identify these different targets on every acquired image, which requires an approximate knowledge of the measured object.

4.2.3 Cases of use

The specificities of photogrammetric metrology must be looked for in relation to other methods of measure, in order to identify its typical uses. We may consider it especially useful in the following domains:

- when the measure must be acquired in a very short time on numerous to very numerous points (from tens to hundreds of thousands of points);
- when measures are to be performed without possible or desirable contact (very hot objects, contaminated objects, objects for which no pose of targets is admissible);
- when measures can be exploited only a long time after the acquisition of the images: the image acquisition is an inexpensive part of the complete photogrammetric operation, and image processes can be put back to a later period, for example to measure the ageing of a given object.

We will examine here some examples concerning one or several of these domains.

Medical domain

Two situations are typical of this domain: the follow-up geometric of a portion of the human body (lesions, tumours, dermatological problems, but also identification of people by morphometric measures . . .), and the surgical intervention from afar allowing a specialist to intervene quickly on distant sites without displacing the patient or the surgeon. The second one is again in development, but it is likely that it will be limited to transmitting some stereoscopic images from a site equipped for a telemanipulation of surgical instruments, and that photogrammetric aspects will be quite reduced (possible 3D measures from afar). The first requires more classic concepts. Being about a living surface, it will be necessary to proceed to an instantaneous measure of two images forming a stereoscopic couple. And as it is not simple to put in the field of every image some elements permitting a formation of the model in differed time, the most logical answer is to use two cameras fixed very rigidly in relation to each other. Thus, once calibration measures have been achieved, one is able to measure in 3D the part of the body appearing in the zone covered in stereoscopy. An aid by automatic correlation gives assistance in pointing for non-photogrammetres practitioners. However, as we will show in §4.2.4, the extremely low variable radiometry of human skin requires a special lighting so that the whole functions correctly.

Objects in movement

This is quite a tyical case for the use of photogrammetry. We are in the presence of an object that distorts itself quickly, or of an object in movement, and the number of points to measure is superior to a few units (if the number is very low it can be performed with motorized automatic pointing theodolites). Examples of such situations are numerous: study of pieces distorting themselves under strain (tests of plane wings), dynamic phenomenon studies (survey of the wave formation to the stem of a ship, fall of blocks in an unsteady slope), etc. The images are in this case necessarily acquired with several cameras, which must be synchronized rigorously. The geometric elements allowing the setting up to be achieved are obtained according to what is feasible, every situation forming a special case: some known coordinate targets are in the field of cameras, or again localization and orientation in the space of cameras are obtained by external means (GPS, inertial systems, goniometric orientation . . .).

Controls of coppercraft pieces

Pieces of thick coppercraft cause subtle metrological problems: it is generally about the control of the obtained shape in compliance with the initial specifications. Some pieces of very large size can require that the reality be very close to the surfaces calculated (propellers of ships, wings of plane, parabolic reflectors for radio waves for telecommunications, radio-astronomy, etc.). The most economic solution here is still photogrammetry, which permits the requested precision to be reached (often better than 1 mm) in a much more cost-effective way than other methods. The other main technical solution consists in using some tridimensional measuring machines, but these machines are limited in accessible measurements (the largest ones don't permit one to make measurements of objects larger than 3 m), and are slow (several seconds between every measured point), but are on the other hand much more precise (they measure with a precision of the order of the μm). Besides, they are extremely costly, and pieces must be brought to the machine, and not the reverse. Photogrammetry, although not so precise, does not present these limitations. Besides, the acquisition of the whole image can take place in a very short time, as we already saw. Without reaching the complete simultaneity, which requires arranging numerous cameras, it is possible to acquire the whole image if need be in less than one hour. This type of possibility is profitable in the case of large-sized objects for which the temperature creates distortions, that cannot be controlled (objects installed outside, as for parabolic antennas, ship sections intended to be fixed together, etc.).

Surveys of painted underground caves

We are here in the case of artistic productions, onto which it is sometimes completely impossible to put targets considering the fragility of the object. Photogrammetry permits measures without contact, without any risk of destruction for the studied object. In counterpart, one sometimes expects in such cases the possibility of restoring the acquired images under a plane shape, or developed on a mean plane, which is evidently not possible without the calculation of a complete 3D model of this object. So a fresco achieved on a surface that includes numerous mouldings, or tracings of animals on prehistoric underground caves, are cases in which elements of reliefs of the cave walls may be an integral part of the tracing and are therefore indispensable for the archaeologists' interpretation. In such situations, it is necessary to try to identify some precise and punctual details in sufficient quantities that will be determined by classic topometric methods. If this is not possible, one can create some relatively precise temporary details while using low-power visible laser beams (red laser diode pointers, or impacts of reflectorless laser EDMs, of Disto type (Leica) for example). There is also the possibility of working with phototheodolites, which can be used in order to deliver oriented and localized images with a high precision. It allows images to be restored without using elements of location in the images themselves. The inconvenience of this solution resides in the impossibility of the operator guaranteeing himself against a lawlessness of the material under utilization. For example, if the optic axes of the theodolite and the camera have relative directions that change without the knowledge of the operator (shocks, vibrations), it will a posteriori induce mistakes that one will often not be able to correct. It is then veritably a blind work.

Surveys of high-temperature objects

The possibility of measuring with absolutely no contact with the measured object is therefore a classic field of the application of photogrammetry. So to control streams of incandescent metal, there is virtually no equivalent solution. Cameras are then adapted to such environments, but processes done on images are always of photogrammetric type. One benefits then from the capacity to measure without contact, as well as the one to acquire images instantaneously. Two stationary, rigid cameras can be installed permanently, and their relative positions will be the topics of a complete calibration by taking images of a tridimensional object including targets, whose relative positions will have been measured with a very high precision.

4.2.4 Typical problems of photogrammetric metrology

Photogrammetric metrology may be adapted, as a question of principle, to situations that change often, and at each time it must be adjusted to the different cases, as we have seen previously. Nevertheless, certain technical problems are recurrent, because they are bound to the same bases of the photogrammetric process. We will present here certain among them.

Case of uniform surfaces

This case is often encountered: a uniform surface will be defined here as a surface whose apparent radiometry varies very little on an important area. This very weak variation itself is a function of two parameters: the dynamics of the sensor, and the real variation of the radiometry. The reason why this dynamics is very important is it becomes possible to measure some extremely reduced radiometry differences, and thus the uniform zones become rarer and in any case of more reduced size than with a more modest dynamics sensor. Surfaces of nearly uniform radiometry are very frequent, and are met with in most of the previously evoked cases. So in medical imagery, they are surfaces of skin, or in coppercraft, surfaces of raw metal. On such surfaces it is then impossible to point with the eye, and tools of automatic correlation may not be able to provide some satisfactory indications. One solution may prove to be satisfactory in such cases: it consists in using a special lighting if the situation permits it. This lighting is chosen in order to project a whole set of random spots on the object. For example, for a metric dimension object one will be able to use a lighting provided by a retroprojector, on the tray of which will have been arranged a film producing such very tight random spots. Of course it will be necessary that the intended motive does not include any periodicity, otherwise it will become very difficult to avoid false correlations.

Case of reflecting surfaces

In the same way it will be impossible to process correctly by photogrammetry reflecting surfaces without making a particular preparation of the object to process. The solution proposed most frequently consists in such cases (polite metal, glazing, etc.) of putting down a thin talc deposit, and there again illuminating the surface while projecting some random spots. The deposit of talc is easy to create, and doesn't present any difficulties being removed. But this method will not be capable of being put into operation in some situations, and it is necessary to know how to anticipate results then that will be locally mediocre (for example, liquid surface cases, or again humid surfaces in decorated underground caves).

Numbering of targets

In most areas of photogrammetry, and it is of course the case in those of metrology for which one searches for the best possible precision, one uses targets of measurements adapted to the acquired images. These targets serve to do the aerotriangulation on the one hand (even though the term is not suited to cases of lack of aerial images, it remains used in spite of everything) and the setting up of couples. On the other hand, targets may be necessary as being points to be determined with the best precision: the pointing on targets are indeed more precise than pointing on natural surfaces. Some studies thus require a very large number of targets, and the problem is that it is necessary to identify these targets, in order to be able to find without error from one image to the other what are the homologous points. When the object to survey is very large (architectural photogrammetry, for example), the number of each target can be marked by hand with chalk. But on objects of some metres, and especially if targets are very numerous, this principle cannot be kept. One can then use labels including bar codes, capable of being read automatically at the time of the automated process of images. Nevertheless, this solution brings strong strains on the orientation of these labels, as they can become illegible when they are seen too much in perspective. When one uses classic reflecting targets and an annular flash mentioned previously, it becomes possible to localize targets automatically in the image by the use of a simple threshold. In such a case, if one provides the software with the approximate geometry of the object on which are the targets, it can itself then proceed to a numbering of targets, which allows a considerable time to be saved at the time of the exploitation of images.

4.2.5 Findings, perspectives

Photogrammetric metrology is a part of metrology that has still golden days ahead. It permits some types of works to be processed in a much more cost-effective way than any other methodologies. Besides, the appearance of direct CCD digital images in the 1990s, opened the doors of the real-time process. It must not be considered as a competitor with other techniques (utilization of reflectorless EDMs, of laser scanners systems, of tridimensional measuring machines). It is, for example, faster and more precise that the first, slower, less exhaustive and much more precise that the second, and for the third it is without contact, far less precise and considerably faster. It can be adjusted to objects of all sizes, and one of its essential steps (the image acquisitions) permits a storage that will allow all the processes to be repeated if need be. It allows one not to process images, therefore, and to keep them a very long time for possible future processes. Finally, let us wager that this discipline will certainly progress again, since it entered the digital era quite late. For example, one can expect to see phototheodolites used more and more in metrology, to provide measures of distortions of distant objects from afar.

The inter-techniques collaboration between photogrammetry, image processing and robotics will probably provide many new tools in the next few years.

Index

adaptive filtering 93
adaptive sampling meshes 162
additive synthesis 27
aerotriangulation 6, 8, 12, 15, 62, 63, 78, 116, 118, 119, 120, 123, 124, 125, 126, 131, 132, 134, 135, 139, 140, 149, 156, 157, 287, 299, 346
airborne laser ranging systems (ALRS) 53, 54, 55, 56, 57, 58
ambiguities: matching 170, 207; resolution of on flight 117
angular fields of view 34
atmospheric diffusion 16, 21, 22, 35
atmospheric refraction 1, 8, 12, 49
attenuation 17, 19, 20, 21, 22

basins 255, 268, 269
bundle restitution 304, 305

catchment areas 258, 264, 268, 270, 276, 277, 278, 281, 282
CCD 1, 14, 23, 24, 34, 36, 37, 38, 39, 41, 42, 43, 60, 61, 63, 66, 76, 153, 168, 288, 300, 307, 311, 312, 313, 314, 340, 341, 346
characteristic features 251
characteristic lines 164, 253, 255, 256, 257, 259, 271, 275, 277, 278, 294; networks 280
CIPA (International Committee for Architectural Photogrammetry) 300, 315, 334, 335, 336, 337, 338, 339
close range photogrammetry 337
collinearity 3, 13, 15, 124, 148, 155, 340; equation 3, 13, 124, 155
colorimetric spaces 26, 27, 32
compensation of forward motion 24

compression: algorithm 101, 105; by fractals 114; of images 100; rates of 101, 104–6, 109, 114; reversible 101, 102, 103, 106; with losses 101, 104, 108
constructive solid geometry 330, 331
contour lines 159, 160, 164, 229, 230, 232, 233, 234, 239, 250, 256, 257, 266, 272, 277, 292
convolution operators 89
correlation templates 184
crests 234, 255, 256, 257, 258, 271, 278, 279, 280
cylindro-conical geometry 38

decorrelators 105, 108
Delaunay triangulation 162, 225, 229, 233, 235, 251
diffusion by sprays 17, 18
digital cameras 24, 41, 45, 56, 64, 67, 119, 124, 153, 158, 288, 294, 299, 311, 312, 314, 316
digital elevation models (DEM) 78, 159, 160, 168, 169, 181, 195, 197, 202, 208, 220, 224, 225, 226, 281, 282
digital photogrammetric workstations (DPW) 59, 78, 145, 156, 176, 224, 300, 314, 323–4, 338
digital surface models (DSM) 43, 47, 49, 50, 53, 78, 159, 160, 162, 164–7, 194–6, 202, 210–25, 250, 288, 289, 301, 322, 323, 339
digital terrain models (DTM) 38, 46, 52, 55–8, 62, 78, 118, 131, 152, 156–60, 166, 168, 170, 177, 207, 212, 220–3, 229, 251, 255, 258, 261, 263–71, 277–87, 289, 294, 299, 314, 317

Index

digitization 39, 43, 58, 59, 60, 63, 64–5, 68, 76, 100, 152, 158, 164, 209, 253, 308, 310, 311; of aerial pictures 58
discontinuities 29, 37, 61, 81, 162, 187, 190, 198, 200, 202, 211, 214, 216, 225, 254, 271, 280
disparity 125, 131, 137, 141, 177, 189, 192, 196, 200, 204, 205
distortion 1, 6, 12, 13, 14, 15, 23, 24, 37, 45, 63, 101, 104, 105, 108, 112, 114, 173, 176, 287, 302, 303, 319, 341
drainage networks 257, 258, 277, 281
dynamic programming 200, 202, 203, 205, 207
dynamics 43, 44, 46, 56, 57, 65, 76, 79, 81, 82, 85, 153, 340, 345

Earth curvature 7
elastic grid surfaces 231, 243, 246
entropy 103, 104, 105
epipolar lines 151, 174, 175, 180, 200, 202, 203, 205
epipolar resampling 148, 150, 152, 175, 176
equalization of histograms 82

false matches 188, 190, 216
field curvature 23
filtering 88, 89, 91, 93, 95, 99, 100, 111, 112, 131, 132, 141, 143, 156, 190, 217, 218, 219, 221, 224, 271, 279
fog 16, 17, 22, 23, 41, 44, 46
forward motion compensation (FMC) 36, 43, 60

GALILEO 115, 124
Gaussian filters 91
geographic information systems (GIS) 59, 155, 156, 220, 278, 285
geoids 7
geometric dilution of precision (GDOP) 122
glass plates 12, 14, 15, 62, 341
global positioning system (GPS) 8, 36, 37, 38, 42, 53, 54, 55, 56, 78, 115, 116, 117, 118, 120, 121, 122, 123, 124, 156, 285, 287, 343
GLONASS (see GALILEO)
GNSS 115, 124

hidden parts 15, 45, 46, 57, 190, 208, 222
histograms of images 79; manipulations of 81
homologous points 125, 128, 129, 131, 132, 134, 137, 139, 140, 141, 142, 143, 152, 170, 171, 173, 175, 178, 179, 185, 187, 198, 202, 205, 318, 322, 346
homomorphic filtering 88, 95
hot spots 34, 172, 289, 291, 334
HSV (hue, saturation, intensity value) space 27, 28, 29, 30
Huffman code 106, 107

ICOMOS 334, 335, 339
image pyramids 150, 320
index map 78, 124, 140, 142, 143
inertial measurement units (IMU) 118
interferogram 49, 50
interferometry 47, 48, 49, 50, 52, 53, 120, 164, 165

join-up lines 289
JPEG 102, 106, 108, 109, 112, 114, 153

Karhunen-Loève transform (KLT) 113
kinematics 118

lasers 1, 53, 54, 58, 118, 120, 164, 224, 225, 228, 289, 344, 346
linear filtering 89, 91

median filters 93, 97
meshes 160, 166, 255, 260, 261, 262, 263, 264, 267, 272, 281
Mie diffusions 16, 17, 18
multi-image texture and radiometric similarity (MITRAS) 214
multi-image texture similarity (MITS) 213, 214
multi-image similarity function 212
multiplicity 96, 131

navigation 36, 54, 56, 115, 116, 118
noise 1, 22, 43, 50, 61, 62, 64, 65, 66, 68, 70, 71, 73, 76, 87, 88, 89, 91, 93, 94, 96, 97, 98, 99, 100, 103, 112, 116, 123, 136, 157, 167, 170, 183, 189, 208, 209, 211, 221, 225, 266, 312

official national reference frame 285
orthoimages 212, 217, 289, 303
orthophotography 41, 46, 56, 59, 62, 118, 158–9, 221, 222, 223, 282, 283, 284, 285, 286, 288, 290, 291, 292, 299
ortho-templates 184, 192, 197
OTF (on the fly) 117, 122

passages 29, 33, 98, 140, 255, 256, 259, 269, 271
perspective 2, 6, 13, 14, 42, 43, 115, 116, 117, 120, 121, 131, 136, 137, 148, 157, 184, 203, 282, 316, 317, 333, 346
perspective centres 115, 116, 121
Photo-CD system 311
photogrammetric filtering 131
planarity of emulsions 14
points of interest 125, 127
pressurized planes 12

radar 1, 47, 48, 49, 50, 51, 52, 53, 120, 148, 165
radarclinometry 47, 48, 50, 51
radargrammetry 47, 48, 49, 50
radiance 17, 21, 22, 25, 34
radiometric balancing 288, 290
radiometric quality 62, 63, 64, 66, 73, 76, 77, 289, 310
raised structures 159, 202, 220, 221, 222, 224, 294
raster grids (RG) 160, 161, 212
Rayleigh diffusion 17, 18, 20, 22
reference atmosphere 9
refraction angles 10
rotation matrix 2, 4, 5

scanners 1, 38, 58, 59, 60, 61, 63, 288, 310, 311, 314, 336, 346
segmentation 34, 73, 221, 222, 223, 224
semi-metric cameras 308, 309, 310
shape from shading 48, 50, 52, 164, 165
shutters 24, 36, 121
similarity measures 125, 127, 169, 207
spectral bands 25, 33, 34, 112

spectral decorrelation 112
stereo matching from image space (SMI) 173, 183, 184, 186, 189–90, 192, 195, 198, 202, 211, 220, 251, 319
stereo matching from object space (SMO) 183, 189, 190, 192, 195, 197, 200
stereopreparation 118, 123, 287
subpixel displacement 149
subtractive synthesis 27
summits 2, 266, 281
surface models 159, 303, 305, 328, 329, 330, 331
systematism 15, 66, 123, 124, 127, 257

terrestrial photogrammetry 10, 11, 14, 137, 162, 300
thalwegs 234, 239, 253, 255, 257, 258, 259, 260, 264, 265, 266, 267, 269, 270, 272, 275, 277, 278, 281, 282
thin plate spline surfaces 231, 236, 239, 240, 243
tie points 119, 124, 125, 126, 127, 130, 131, 132, 134, 135, 136, 140, 144, 156, 196
time delay integration (TDI) 24, 60
topographic surfaces 166, 167, 229, 231, 232, 233, 235, 236, 239, 243, 254, 257, 258, 266, 281
trajectography 38, 42, 116, 118, 119, 122
triangular irregular networks (TIN) 160, 162–4, 168, 196, 225–6, 231–2, 235, 251
trichromatic vision 26

vignettage 23, 45
visibility 19, 20, 21, 73, 122, 198, 289
volumetric models 326
VRML 325, 333

watersheds 255, 257
wavelet transforms 105, 108, 110, 111, 112, 113
wireframe models 326, 327, 328